the BackYard Orchardist

A complete guide to growing fruit trees in the home garden

p 26

1) broadly define a grp of fruit trees
suitable to our area —

narrow down choices best for our particular situation
micro climate

D0103665

the Backyard Orchardist:
A complete guide to growing fruit trees in the home garden

Stella Otto's first book is:
* Winner - 1994 Benjamin Franklin Award for Best First Book
* Finalist for Best Garden Book 1994 Benjamin Franklin Award
* Garden Book Club Alternate Selection
* Rodale Organic Gardening Book Club Alternate Selection

Here's what reviewers have to say:

"*The Backyard Orchardist* is a first rate effort and will fill a conspicuous void on the bookshelf."
     Horticulture

"This little gem is the finest single source of fruit growing information published to date....very comprehensive..."
     Pomona, newsletter of the North American Fruit Explorers

"....packed with down-to-earth information that the home gardener and master gardener crave."
     Journal of Small Fruit and Viticulture

"What we like most about this book is it isn't intimidating....she gently and convincingly enables us to realize that, yes, *we can grow fruit successfully*.
....bulging with facts which can be put to work by both novice and experienced amateur fruit growers."
     HortIdeas

"...the finest reference (by far) for the beginner!"
     Ed Fackler, Rocky Meadow Orchard & Nursery

"...If you're still daydreaming about that backyard orchard, there are plenty of trees to pick from out there in garden land. But you'd be wise to buy a copy of *the Backyard Orchardist* and get smart first."
     Huntsville Times

In praise of
the Backyard Berry Book
A hands-on guide to growing berries, brambles, &
vine fruit in the home garden
by Stella Otto

"written in plain hands-on language, clear and purposeful drawings, *the* definitive how-to guide to small fruit gardening."
BackHome

"...enjoyable to read, easy to refer to."
Fruit Gardener

"Stella Otto has done it again! When we reviewed her first book, *the Backyard Orchardist*, we wrote "What we like most about this book is that it isn't intimidating. Stella Otto doesn't act like some expert-from-on high; she gently and convincingly enables us to realize that, yes, we can grow fruit successfully." Ditto for *the Backyard Berry Book*.....you should by all means read this book."
HortIdeas

"likely to be much-thumbed as a reference by the green-thumbed crowd."
Small Press

"one of the best books on growing berries I've seen in a long time."
Newark (NJ) Star-Ledger

◆ By Benjamin Franklin Award-winning author
◆ Garden Book Club Alternate Selection
◆ Rodale Organic Gardening Book Club Alternate Selection
◆ Feature excerpt in *Country Journal* magazine
◆ Reviews in newspapers, magazines and trade publications including American Homestyle & Garden, Better Homes & Gardens Deck and Landscape Planner, Booklist, and Bloomsbury Review

the BackYard Orchardist

A complete guide to growing fruit trees in the home garden

by Stella Otto

OttoGraphics • *Maple City, Michigan*

Publisher's Cataloging-in-Publication Data

Otto, Stella B.
    the Backyard Orchardist: A complete guide to growing fruit trees in the home garden/by Stella Otto.
        p. cm.
    Includes index.
1. Fruit-culture. 2. Fruit-tree fruit. 3.Fruit-varieties. I. Title
SB355.O8   1993   634-dc20   92-96980
ISBN 0-9634520-3-7

Foreign rights and specialty sales please contact publisher:
OttoGraphics, 8082 Maple City Rd., Maple City, MI 49664.
Phone: 616-228-7513 or after June 1, 1999: 231-228-7513.

Trade distribution by:
Chelsea Green Publishing Co.
P.O. Box 428, White River Junction, VT 05001.
Phone: 802-295-6300, Fax: 802-295-6444
To Order: 1-800-639-4099, Fax: 603-448-2576.

Manufactured in the United States of America

10 9 8

Contents

SECTION I. Getting Started with Fruit Trees

SECTION II. Fruit Fundamentals - Growth Habits of Specific Tree Fruit

Stone Fruit

SECTION III. Caring for Your Fruit Tree

List of Illustrations

List of Charts

Quick Reference Charts

Acknowledgments

Over the years of fruit farming, and now in writing, I have been fortunate to have had the support and encouragement of many fine people. I would like to extend my deepest gratitude to a number of them. Without their help, this book would not have been possible.

Special appreciation, beyond words, goes to Karen Becker and Ellen van Gemeren for the many hours spent editing, helping with clarifications, and proofreading. To Larry Mawby, whose unique creative sense made this work both more inviting and more useable, I owe enormous thanks. Larry, you are one of the people who made this project truly fun and I would like to really encourage you to pursue your writing project - I think you'll find it will be more pleasure than work!

I give sincere thanks also to A. J. Bullard, Tanya Denkla, Julie Francke, Sucile Mellor, Cathy Carter, Kate Bandos, Katie Shaw, Skip and Betty Osborne, Sally Berlin, Ed Fackler, Robert Nitschke, Tom Vorbeck, Jim Nugent, Jim Johnson, Andy Norman, Dave Rosenburger, Sarah Wolfgang, Michael Willet, Tara Baugher, Jim Bardenhagen, Mary McClellen, Greg Patchen and Al Kuepfer, who all helped in one way or another; providing critique, moral support, technical information, or an education in the many facets of book publishing. A special thank you also goes to Laurie Davis, for her quiet, but not unnoticed efforts beyond the call of duty, in designing and illustrating the cover.

Last, but very importantly, I am grateful to my husband and son for their patience, as I worked overtime to stay on schedule.

Preface

This book is written for all of you who have thought about including a small planting of fruit trees in your yard or rural acreage, but didn't know where and how to start.

Maybe you think ....

    .... your yard is too small,
    .... your climate is too hot or too cold,
    .... your soil is not suitable,
    .... you don't have enough time...

Throw all those doubts on the compost heap. You too can be a Backyard Orchardist!!! Growing fruit trees successfully is not really that difficult. Even a neophyte gardener can do it. You can too!

In what follows, you will learn how to select a spot for your fruit trees, or select the right tree to fit your spot; how to plant it, feed it, prune it, keep it free of insects and diseases, and finally harvest a crop you will be proud to share with your family, friends, and neighbors.

Writing about all of this was easy. What was not so easy, was deciding what to leave out (or save for a future book).

As you read through the chapters that follow, you may become aware that, unlike the few other available books on fruit growing, the Backyard Orchardist does not take up a lot of space describing individual varieties. In most cases since many good descriptions can be found in nursery catalogs, the varieties are listed by name only. Instead, space is devoted to information that you might not find elsewhere and to more extensive illustrations. Yes there is even a list of nurseries, so you can easily get those catalogs.

Topics that are often intimidating to and not immediately needed by the beginning fruit grower have been saved for future works - propagation, training an espalier tree, biological monitoring. So get out that shovel and start working some soil! Before you know it, the tree you plant now will be bearing fruit and you will be an "old pro" ready to learn about these other exciting facets of fruit growing.

Section I.

Getting Started with Fruit Trees

1. Enjoying Fruit Trees In Your Landscape

The lore and love of fruit trees has been with us for many generations. Johnny Appleseed, pioneer of apple orchards, has long been a folk hero and George Washington brought the cherry tree to fame with his famous "I cannot tell a lie" declaration. For many gardeners the attraction of growing a fruit tree is as strong today as it was during the days of these well known figures.

The wonderful thing about fruit trees is that they can be enjoyed by virtually anyone willing to invest a little bit of energy and patience into growing them and they can be enjoyed in so many different ways. Whether you are a novice gardener who would like to start with something unique or a seasoned veteran looking for a new adventure, fruit trees are rewarding, interesting and relatively simple to maintain once you understand some basics. Over my many years as an orchard owner and consultant, I have been asked many questions; some simple, some complex. From those questions this book has grown. To introduce you to this exciting hobby, *the Backyard Orchardist* will explore the basic concepts of fruit growing and guide you through the skills needed to successfully raise your own fruit trees.

One of the first things you will learn is that fruit growing is an avocation that takes patience and a little bit of forethought. For developing these virtues you will be well rewarded, however. You will have the pride of saying "I grew it myself" and the opportunity to enjoy the sweetness and flavor that only tree ripened fruit offers. You will also become privy to the understanding that is behind the mystique of fruit growing. *The Backyard Orchardist* is meant to be an encouraging hand and personal consultant throughout the adventures of fruit growing. You will soon find out that fruit growing is both a science and an art. As you begin your adventure with fruit trees, you may want to read through *the Backyard Orchardist* completely and then return to those chapters most pertinent to your current needs. You will also come to see that many of the things you do to your fruit tree are interrelated and as you take one action you may save yourself another. So, explore, observe, keep a few notes and don't be afraid to try something different if what you are currently doing is not bringing the expected results. Above all relax, be patient and enjoy! Most fruit trees are quite resilient and will allow you to make a few "mistakes" along the way. Consider them learning experiences and don't worry; most fruit trees will bounce back. Although there is much to be learned, it certainly doesn't all need to be done at once. Start small with one or two of the easily grown trees and by the time they are bearing fruit you may find that you have a much greener thumb than you ever envisioned. Yes, growing fruit trees is fun and not all that difficult. Let's take a look now at the many ways that *you* can enjoy fruit trees in your landscape.

Fruit Trees for Personal Enjoyment

Most gardeners consider feeling the sun on their backs and the soil in their hands pleasure in itself. Much satisfaction comes from nurturing a plant while enjoying the out of doors. Many are the rewards of watching the fruit tree grow and change with the seasons knowing you play a part.

Growing a fruit tree also brings a great deal pride, along with feelings of success and self sufficiency. Being able to share the fruits of your labor with friends and family while saying, "I grew it myself", can be as rewarding as the joy of knowing that you have learned new skills in the process.

For Visual Beauty

Long before your fruit tree provides you with delights of the palate, it will delight the eye. The apple tree's light pink bloom turning to delicate white is a sight to behold. An apricot, peach or nectarine tree with its profuse pink bloom can be a showpiece in your yard! For a real conversation piece, consider an espalier pear tree. You can use a single fruit tree to direct the eye to a focus in your landscape or a collection of trees to fill a void in your yard.

For a Bountiful Harvest

High on most people's list of reasons for wanting to grow a fruit tree is the enjoyment they will get from the fruit itself. There are many ways to enjoy your harvest. Growing your own fruit allows you access to uncommon as well as your favorite varieties. This may be especially important if you yearn for a particular "old fashioned" variety such as you might remember from grampa's farm.

Many of the older, or antique, varieties are not being raised commercially today for a number of reasons. By today's market standards, many have a less than perfect appearance, a short storage life or a tendency to bruise easily that makes them unsuitable for long distance shipping. Since you probably will not be selling your harvest, you won't need to be very concerned with some of these slight commercial limitations. Instead, you can concentrate on enjoying what these old time fruit varieties have to offer -- flavor! Many antique varieties have unique flavors hard to find on today's grocery shelf. Fruit is sweetest and juiciest when picked fully ripe, but at that point it is too perishable to ship to distant markets. Plums and peaches are particularly flavorful, but only the home orchardist has the benefit of enjoying them at their peak. Aside from eating fruit fresh you will also be able to preserve it for use in the off-season. Many of the antique varieties are particularly suited to drying, canning, freezing or turning them into all sorts of wonderful jams, jellies and chutneys.

For Health and Nutrition Benefits

Everywhere we read about the benefits of more exercise and the need to eat more fruit and vegetables for fiber and naturally occurring vitamins. By planting fruit trees in your yard, you are on your way to incorporating both of these healthy practices into your lifestyle almost effortlessly. The regular care that fruit trees require will give you a pleasant way to work a bit of relaxed exercise into your routine year around and the fruit you enjoy from your harvest will be a benefit to your diet. For people concerned about chemicals and pesticides, growing your own fruit can be one way to be more fully aware of what substances are used on the food you eat.

As A Climate Modifier

Properly placed, fruit trees can be used to create small areas of shade to shield a large window from the sun during the warmest times of day or shelter a picnic table for a cool summer meal. In windy areas, several fruit trees planted closely together can act as a nice windbreak to protect your home from winter winds and help keep the heat bill down. While a hedge row of fruit trees can protect against the wind, it can also be useful as a sound muffling barrier if you need protection from noisy neighbors or a busy street.

As Erosion Control

Not only can fruit trees help keep noise, dust, and wind out of your yard, they can keep your garden's soil where it belongs. The root mass of a tree will cover underground roughly the same amount of space that the branches take up above ground. By holding the soil in place with the tangle of roots that continues to expand underground, your fruit tree can help to stop erosion on a hilly site during hard rains, whipping wind or heavy spring thaws. Leaves falling from the tree at the end of each growing season help to act as buffers against driving rains. Additionally, decaying leaves add to the organic matter and structure of the soil, enhancing marginal soils and making them better able to withstand the forces of nature.

As Wildlife Habitat

For many people, a most enjoyable activity is watching the birds that come to their window feeders. Fruit trees can serve as a protective area for shy bird species while they work up the courage to visit your feeder and as nesting places throughout the season. Placing a tree within viewing distance can often allow you to enjoy the sight of a flock of Cedar Waxwings in late winter or numerous Grosbeaks. Both species enjoy eating any leftover apples that may hang from a tree and can be easily encouraged if a share of the crop is left for them. Inviting birds to your yard will also give you the benefit of their appetite for numerous insects and help reduce the population of pests that may bother humans or your fruit crops.

2. Selecting a Site

Deciding where to plant your fruit tree can be one of the most important decisions you make regarding the future and longevity of your home orchard. In this chapter we will take a close look at the conditions in your yard and how they are suited to your fruit trees. As you read through the following sections, you will probably feel that you have a lot to learn and many decisions to make. Don't let this scare you. Every yard has some place that is suitable or can be adapted for a fruit tree.

So many aspects of selecting and growing fruit trees are interrelated, that you may find it worthwhile to read this book through completely and then return to this chapter. As you narrow down your choices of type and variety of trees, rereading this chapter can help you tie together all of the aspects of site selection that will lead to a long and prosperous life for your fruit trees.

A Close Look at Your Site

In deciding on the ideal location, you should take a careful look at the area you have available for planting. This can be done by taking a piece of paper and drawing a reasonably accurate "map" of the site. Start by pacing out the rough dimensions of your property and drawing them on your map. Next, draw in those parts of your landscape that are already permanent: your house, garage, a fence maybe, and any large existing trees or shrubs that you plan to keep. Note unsightly areas that you would like to hide or improve. Also include the location of any water source, be it a pond, stream or water faucet. Then, show areas of differing topography. You may want to use a few colored pencils for this. Pay particular attention to areas that are considerably higher or lower than the surrounding terrain. Now you will have a guide, like the one in Figure 1, on which to note the important growing factors that will be discussed next.

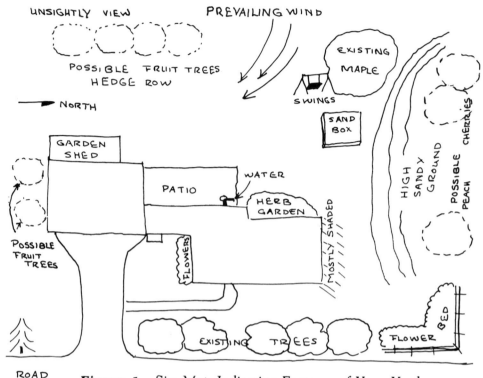

UNSIGHTLY VIEW

PREVAILING WIND

POSSIBLE FRUIT TREES
HEDGE ROW

NORTH

EXISTING
MAPLE

SWINGS

SAND
BOX

GARDEN
SHED

WATER

PATIO

HERB
GARDEN

FLOWERS

POSSIBLE
FRUIT
TREES

MOSTLY SHADED

HIGH SANDY GROUND

POSSIBLE PEACH

POSSIBLE CHERRIES

FLOWER BED

EXISTING TREES

FLOWER BED

ROAD

Figure 1. *Site Map Indicating Features of Your Yard*

Climate

The first major environmental characteristic to consider in regard to your garden plan is climate. This can be broken down into two areas of concern: the normally expected climate in your broad general area and the microclimate established by certain characteristics unique to your property. First, let's take a look at the general climatic requirements for growing fruit trees. This book covers what are commonly considered temperate zone fruit trees. The temperate zone is generally defined to be from 23 degrees latitude to the Arctic Circle, in the northern hemisphere. Most of the geographic regions of the eastern seaboard, the midwest and the west coast of the United States are suitable to fruit growing as long as attention is paid to selecting the appropriate type and variety of fruit (Figure 2). Comparable growing areas are also found in Canada.

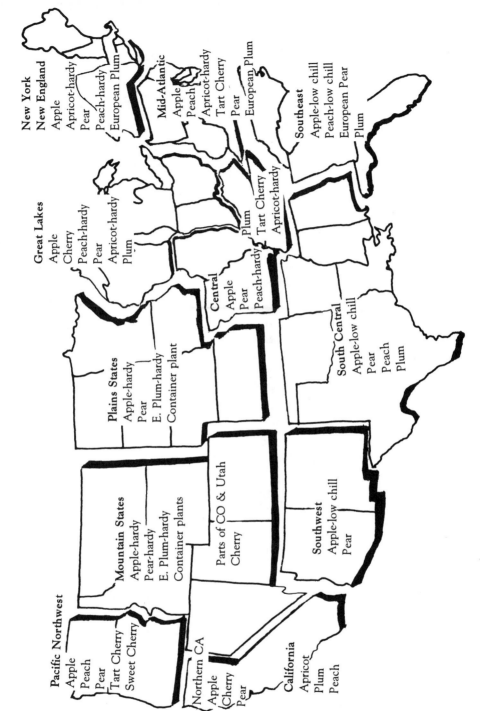

Figure 2. Best Fruit Choices for United States Regions

Average Minimum Temperatures and Hardiness

The two general climatic considerations that must be taken into account are the area's average annual minimum temperature and the average length of its growing season. Most fruit trees can grow in areas with a higher average minimum temperature, but may suffer cold damage in zones much colder than that for which they have been designated. Although fruit trees cannot withstand excessive cold, they do require a minimum number of hours of temperature between 45° F and 32° F to satisfy their dormant rest period needs. Specific chilling requirements of the various types of fruit will be discussed in more depth in Chapter 13. As can be noted from the climate zone map (Figure 3), the warmer areas are generally found in the south and become progressively colder as one moves north.

Another factor that can affect the average minimum temperature is location close to a large body of water. Because large bodies of water tend to cool down more slowly in the fall and warm more slowly in the spring, they minimize temperature extremes and have a stabilizing effect on the fruit production capabilities of the adjacent land within several miles of their shorelines. This is often referred to as a microclimate. Later we will discuss another microclimate, the one found in your yard. Large areas of microclimate are found in Washington, California and Oregon as well as along the west shore of Michigan, New York and Ontario where winds are moderated as they pass over the Pacific ocean or Lakes Michigan and Erie causing a climate that has allowed for development of some of the largest fruit growing regions in North America. Conversely, when the prevailing winds pass over large cold bodies of land the local climate can be particularly harsh and more of a challenge to fruit growing, as is the case in some of the Mountain and Plains states.

When selecting fruit trees for your particular location, be sure to choose varieties rated for your climate zone or colder zones. Fruit trees are normally most hardy in the depth of winter. At that time apples, the most hardy of the fruit trees, can withstand temperatures as low as -25° F. Peaches, on the other hand, can not tolerate temperatures much below -15° F. In order from most to least hardy you would generally find apple, pear, apricot, domestic plum, tart cherry, sweet cherry, Japanese plum, and peach. Most nurseries that sell fruit trees have years of experience and are knowledgeable about which varieties are best suited to different locales. Mail order nursery

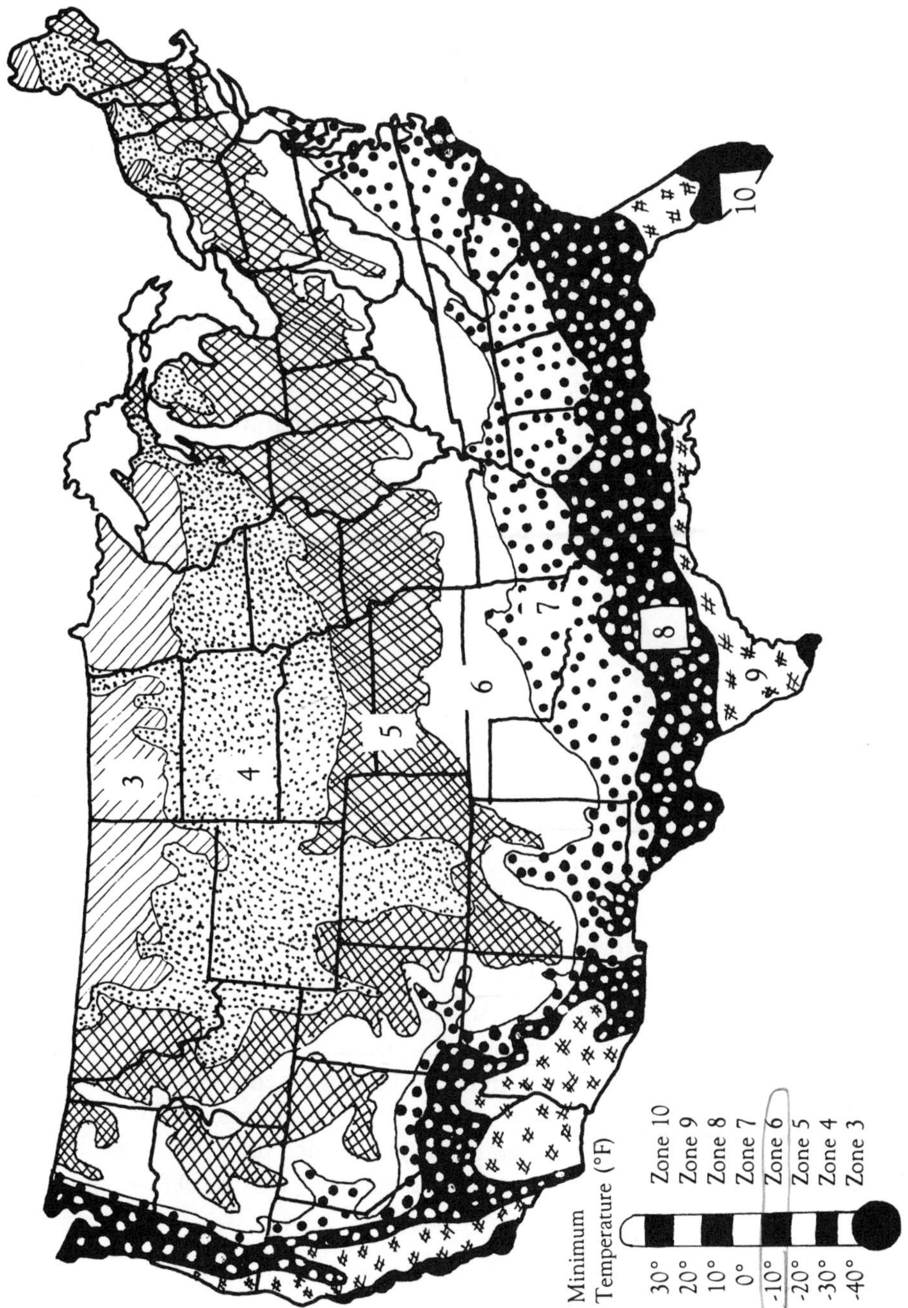
Minimum Temperature (°F)
30° Zone 10
20° Zone 9
10° Zone 8
0° Zone 7
-10° Zone 6
-20° Zone 5
-30° Zone 4
-40° Zone 3
3
4
5
6
7
8
9
10

Figure 3. *United States Hardiness Zones*

personnel can often be very helpful in recommending which of their trees would be best suited to your needs. Most local garden centers stock those trees that grow best in their particular area, so buying from a reputable local garden center can be another way to assure a proper selection.

Length of Growing Season ~~# Frost~~ "or Free days"

The length of the growing season in your area also has an effect on which type and variety of fruit trees would be most productive for you. "Frost Free Days" is a term often used to define the length of the growing season in a given area. It is handy information to have so that you can choose varieties that will bloom, produce fruit and be harvested without damage from major frost. Fortunately, a record of frost free days has been statistically determined over many years of record keeping by various government agencies and will give as accurate information as you need at this time. Refer to Figure 4, which give days below 28° F. This is generally suitable for determining frost free days for tree fruit in your area.

"Growing Degree Days" is another piece of information of use to those who would like to be more specific about a location's growing season. It is merely a measure of accumulated heat units above the temperature at which plant and insect growth effectively becomes active (often referred to as the "base" temperature). Most often you will see degree days referred to as calculated from base 42 or base 50.

At this point it is necessary to be aware of degree days only because certain varieties of fruit require a longer growing season, and consequently more degree days, in which to mature than others. Selecting those that are suitable for your area will lead to the most rewarding harvest. More information about the expected growing degree days in your particular area can often be obtained from your Cooperative Extension Service. Each state, and in most areas each county seat, will have an extension service office. You will find them quite helpful with many of your growing and gardening questions.

Microclimate

Now let's return to the map you made earlier. Once you have broadly defined a group of fruit tree varieties that might be suitable to your area, you can use your map to narrow down the choices to

26 the Backyard Orchardist

cooperative Extension Service
ask about Frost free days / Degree days

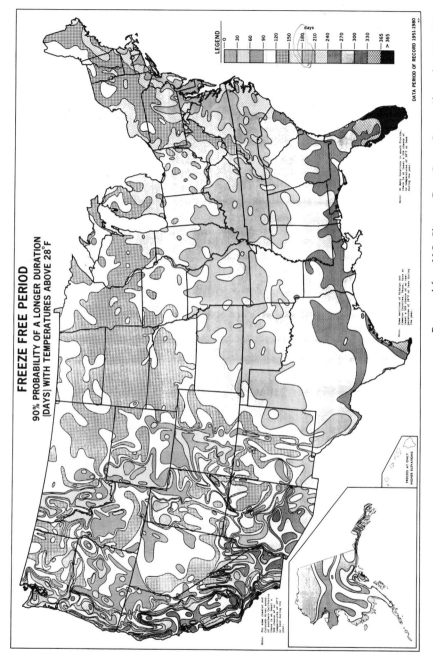

Figure 4. *Average Length of Frost Free Period (days)*

Reprinted from U.S. Climatic Data Center Report Climatology No. 20

Selecting A Site 27

those that will be best for your particular situation. Your map should show major prevailing wind directions, sheltered spots, any tall thick hedge rows of existing trees and areas of higher and lower elevation (if not, you can add them in now). All of these factors affect your yard's microclimate. As already discussed, microclimates are usually caused by unique features of geography. Just as the country and the continent have various microclimate areas, your yard is likely to also.

The most obvious microclimates in your yard will probably be variations in elevation. As you may notice in early spring and late fall, areas of lowest elevation in your yard are prone to be the most frosty. Conversely, high areas are most likely to escape the frost. Look carefully at the surrounding topography as well. What may appear to be a high spot in your yard may still be very low relative to the surrounding area. If your yard sits in a bowl surrounded by higher ground, cold air will collect at the bottom of the bowl and you should select the later flowering fruits. This is important to be aware of; your fruit trees' blossoms, and ultimately the size of your fruit crop, will be influenced by how successfully it avoids being damaged by spring frosts. How completely some varieties mature for harvest will also be influenced in some years by how early they are subjected to hard frosts in the fall.

Microclimates can be influenced by several other factors on your property. One major factor can be your house. Often locating a tree close to the house, especially when it is on a side protected from the wind, can give the effect of being in a warmer microclimate since the house gives off a certain amount of heat. Also, if the tree is placed on the south side of a building where it is sheltered, but exposed to the most direct and intense sunlight, it may act as if it were growing in a warmer microclimate. In a northern climate this can be an advantage by providing a bit more heat for the tree. In the warm southern areas, however, this may prove to be more heat than the tree is able to withstand.

Large hedge rows of tall trees in the vicinity of your planting site will have an effect as well; they may help shelter your yard from strong winds, but they also inhibit air movement. This can lead to cold frosty areas or "frost pockets" on the hedge row's leeward side. In mapping your yard, take this situation into account, even though the tall trees may actually be on your neighbor's property. Take one final look at your site and include any hedge rows on your site map.

28 the Backyard Orchardist

Soil

When you have made a general assessment of the climatic conditions your fruit trees may encounter in your location, the next vital ingredient to consider is the soil. There are several factors to examine when studying your soil and its productive capacity.

First, a good look at your soil texture is in order. If your planting area is flat and small, say a few hundred square feet, very likely your soil type and texture will be fairly uniform throughout. If, however, your area possesses some obvious changes in elevation or covers several acres, you may encounter several different soil types. Each one may be suitable to growing different things. The major soil types can be generally defined as sand, loam, and clay.

Sand, as most of us are familiar with it, is made up of fairly large particles, relatively speaking. These are weathered pieces of minerals and still fairly coarse in texture. Large pore spaces for water and oxygen are found between sand particles. Relatively little organic matter is present in a very sandy soil. Water tends to run through sand quite quickly and sand can also be quite subject to erosion by wind or rain. One place to watch out for sand is right around the foundation of your house. The sand builders use as backfill to provide good foundation drainage is particularly lacking in plant nutrients.

Loam soils are composed of medium size particles. They are a combination of about 50% sand, up to 25% clay, and silt as the remaining soil particles. These soils are commonly found in areas of woodland or pasture where there have been many years of ongoing decay of old leaves, grasses and other plant matter through natural processes. Well aged compost could also be another source of loamy soil. The texture is finer than sand, the soil holds together as a more cohesive mass and has a substantially greater capacity to hold water and nutrients than sand does.

Clay soils are made up of very small, fine textured, highly weathered particles. Clay tends to have very high moisture holding capacity which can sometimes become a problem by waterlogging root areas. Clay particles have strong ionic charges and are chemically attracted to each other. As a result clay soil has a tendency to become very hard, almost like cement, if allowed to dry out completely. This happens because its very small particles fit so closely together. Figure 5 compares the relative soil particle's size.

In between these three main soil types are "combination soils";

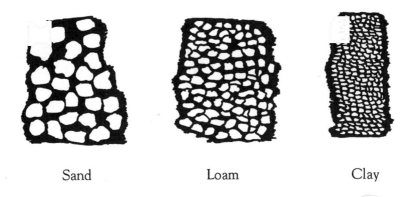

Sand	Loam	Clay

Figure 5. *Magnified View of Sand, Loam, and Clay*

the sandy loams and the clay loams. As their names imply, they are a mixture of a varying percentage of sand and loam or loam and clay. These are often the best soils for fruit trees as they normally have a better balance of desirable characteristics. For example, a sandy loam may have good soil drainage, thanks to the sand portion, and thus will not waterlog root areas. Yet it will also have good ability to hold moisture due to the organic component of the loam. Conversely, clay loams may have high water holding capacity, but do not cement as badly when they dry out because of the air space that is present between the larger particles of organic matter that make up the loam portion. These air spaces will also help prevent the soil from becoming badly water logged.

All other factors being equal, a sandy loam will be the most versatile for growing a selection of different fruit trees. As you will see in upcoming chapters though, there are a number of rootstock choices and soil modifying techniques that will allow you to grow a fruit tree in your particular garden soil.

Soil pH

Here it is - soil pH. Every gardening book talks about it, but most readers are still confused by soil pH. It always sounds like chemistry and quite frankly, it is. For the home gardener, it is also not essential to understand all the chemistry involved with pH measurement. To keep the explanation simple, we will stick to the basic ideas you need to understand in growing your fruit trees.

The term "pH" is actually an abbreviation for the phrase "potential hydrogen" or a measure of the amount of available hydrogen ions in a sample of soil. This measure is used to indicate the acidity or alkalinity (sometimes casually referred to as sweetness) of the soil. Knowing the acidity or alkalinity of a soil can be important because it affects the amount and form of nutrients available to the plant. This will be discussed in more detail in Chapter 10 when we examine fertilizers.

A scale of 0 to 14 is used to express soil pH. To be suitable for fruit trees, it has been generally accepted that a soil should have a pH of 6.0 to 7.0, with 6.5 to 6.8 preferred (although at least one fruit crop not discussed in this book, the blueberry, prefers a pH of 4.5 to 5.5). Most garden centers sell small home pH measurement kits, that although not extremely accurate, can give you a "ball park" pH reading. Another resource for soil testing is your local cooperative extension service which can, for a small fee, send your soil sample to a university lab for testing. It is a good idea to take a complete soil sample *before* starting your backyard orchard (or other permanent planting for that matter), so that you know what you are working with when you start. For a backyard orchard it will be unlikely that you would need to repeat this test very often after that, assuming that your initial pH was in acceptable range and you don't encounter nutrition related problems with the growth of your trees.

Moisture

When making decisions about what to plant and where, other items to keep in mind are moisture and the availability of water. Indeed, this should really be broken down into these two categories, because the presence of moisture does not address only the tree's need for water.

The presence of constantly wet soil on a portion of your property may mean that you have a high water table or poor soil drainage that will make it more difficult to properly grow a fruit tree in that location. The roots of a fruit tree do need a certain amount of water to keep from drying out, but too much of a good thing can raise havoc, too. If a fruit tree's roots are constantly exposed to standing water, it is unlikely that the tree will last long. To test for standing water, simply dig a hole several feet deep at you planting site. Fill it with water and see how fast it drains out. If water is still standing

Selecting A Site   31

after twenty four hours, it would be best to choose another site. Just as the tree needs water, it also needs oxygen to carry on some of its life processes. Some of this oxygen comes to it through the air spaces, or pores, in the soil. If the pores are constantly full of water, as might be the case in a marsh, a bog, or a very heavy clay soil, insufficient oxygen reaches the tree roots and they cannot function properly for nutrient transport. So again, it is important to know your soil type as it may have a bearing on how much moisture a certain area will hold and its suitability to growing fruit trees.

Moisture can also be represented as humidity in the air. In most parts of the country this does not pose a significant problem, but in certain areas where humidity is constantly high, such as the Mid-Atlantic and South East states, it can have an effect on the growth of certain fruit tree diseases. Particularly, many of the fungus diseases thrive under conditions of high humidity and it may be wise to select varieties that are disease resistant if high humidity is a regular occurrence in your area.

Planting trees where they may be subjected to salt spray, such as along the ocean, or more likely from road salt that is splashed into your yard by passing cars during the winter months, can be very detrimental. If you have a salt spray problem, you probably already know it, as your lawn may have dead areas near the road. Certainly avoid planting your fruit trees near these areas, as both the foliage and eventually the spreading roots will be badly affected.

Obviously your fruit tree will need a certain amount of water to grow and prosper. If you are fortunate to live where just the right amount of rain falls at just the right time, you probably don't need to worry. However, this is rarely the case; so for practical purposes you should think about how much water you will need to provide to your tree and where it will come from. The average size fruit tree needs about an inch of rain a week. Choosing a site with loamy soil of one sort or another will make it easier for you in that you probably will not have to water as often, since this soil will retain water better than sand. When locating your tree you may want to think about where you have a water source located. For most homeowners this will be a backyard faucet. If your property is a little larger, you may have a small stream or pond on one corner. As long as these sources have clean water that is not carrying chemicals (especially herbicides), any of them will work fine for irrigating your tree, whether you use a hose or a bucket to haul some water.

In some areas of the country where water is in short supply, or if you just would prefer to conserve water, you might choose to locate a barrel, five gallon can or plastic bucket where it will catch rain water and then use this to water your trees. As long as you have sufficient rain, this method can supplement other watering. To reduce evaporation, keep the barrel covered during dry periods.

To preserve soil moisture once it has found its way to the soil, mulching works quite well. Two to three inches of coarse bark, hay, straw, or even a layer of black plastic mulch will help preserve soil moisture when spread around the base of the tree trunk in a three to four foot circle. Do not, however, use very fine grass clippings or other materials that tend to mat down when wet as they will cause water logging and a host of disease problems. Mulch should also be kept away from direct contact with the trunk as this can encourage the growth of fungus diseases and harbor rodents in the winter who may feast on the bark of your trees.

Sunlight

While climate and moisture are important growth factors for your fruit tree, sunlight is of equal importance, and for several reasons: first, the tree needs sunlight as a vital energy link in the photosynthesis process. Photosynthesis is the chemical process whereby the plant takes in nutrients and water through the roots and atmospheric gasses through the leaves. Using the energy from sunlight, the tree transforms carbon dioxide and water to the carbohydrates it needs for growth. Certain minimum amounts of sunlight are also needed for the tree to adequately produce flower (and consequently fruit producing) buds. This influence will be examined in more detail in Chapter 13 (Flowering & Fruiting) and Chapter 11 (Pruning).

Studies have shown that the average fruit tree needs a minimum of six to eight hours of sunlight daily in order to grow, blossom and produce fruit properly. Therefore, in selecting the location for your fruit trees you will need to choose a place that receives adequate sun exposure. In looking at your site map you may be able to rule out any areas that are directly north of a tall row of trees, a fence or the north side of a building. These areas tend to be shaded most of the day in the northern hemisphere. Try to observe your particular yard during different seasons of the year (but especially in the spring and summer), to be sure that the site receives sun most of the day.

The direction that your site is facing will also affect the intensity of the sunlight it receives. You may be able to use this to your advantage if you have several potential locations for your fruit trees. Normally a site that faces to the south will warm up earlier in the spring and the ground will dry out faster after spring thaws. You may want to locate your trees on a south facing slope if your soil is heavier, such as a clay, to encourage drying and avoid water logged roots. Conversely, if you have a sandy soil you should be aware that your south facing site may require more frequent watering. Also in the southern regions, these south facing slopes could cause the sunlight to be too intense for your trees.

On the other hand, a north facing slope will be the last to thaw and warm up in the northern growing areas. If you are in a location that is subject to late spring frosts, you may find it helpful to locate your trees on a north facing slope. They will be a little bit slower to resume growth and blooming in the spring. This short delay can sometimes make the difference in escaping the last spring frosts that could threaten the blossoms. In a warm area, a north facing slope may give your tree a little bit of heat relief from the very intense sunlight.

Space Required

By now you may be wondering, "How much space is all this going to take?" Depending on the size of the tree, required space can range from about a ten foot diameter circle for a dwarf apple on M9 rootstock to an area thirty-five feet in diameter for a standard size apple or sweet cherry. Typical space requirements for the most common fruit trees are summarized below:

Distance Required Between Fruit Trees

Apple, standard	30 ft.	Cherry, Sweet	20 ft.
Apple, semidwarf	15 ft.	Cherry, Tart	15 ft.
Apple, dwarf	10 ft.	Apricot	20 ft.
Pear, standard	20 ft.	Peach/Nectarine	15 ft.
Pear, semidwarf	15 ft.	Plum	15 ft.

The Perfect Site

It would be so simple if all gardens provided the ideal site for a fruit tree. With the right selection and attention to a little bit of preplanning, many can come close. If one were to define an ideal site for a fruit tree it would likely be a gently rolling site, with some protection from harsh winter winds and extreme temperatures. A growing season with at least 150 frost free days would allow for a wide choice of fruit varieties.

Sandy loam or loam type soils with a pH range of 6.0 to 7.0 would be helpful. Good soil drainage is important. Easy access to water would be beneficial. Top this all off with at least eight hours of sunshine daily.

If this sounds like a spot in your yard, planting time could be near! If you are not sure that your yard is ideal, read on. A number of simple things can be done to turn most any yard into a home orchard. The upcoming chapters will point the way and help you make fruit trees an enjoyable and productive addition to your landscape.

3. Planting and Early Care

Like many enjoyable pursuits, planting a fruit tree takes some preparation. Undoubtedly you are wondering just what to do and where to turn next before bringing that tree home from the nursery or placing your mail order. Ideally, if you can, test and prepare your soil the season before you plant your tree. Most experienced gardeners have learned the virtue of forethought and preplanning. Although you are probably impatient to get your trees in the ground and growing, a little bit of time spent now in proper preparation will pay you back handsomely in healthier, more productive trees and less work in the future.

Preparing the Soil

One of the first steps in preparing for healthy trees is finding out what nutrients your soil can provide. State cooperative extension offices (there is one in each county) and a number of private laboratories provide soil test results for a modest fee. You need to collect a soil sample as described in Chapter 10. Place the sample in a clean plastic bag and bring it to the extension office or send it to the laboratory according to their directions. Ask that the soil be tested for nitrogen, phosphorus, potassium, and soil pH. If your garden has had problems growing other plants, it may be wise to ask for a test of micronutrients and nematodes also. You should get the results back in about four to six weeks.

Meanwhile, rototill or use a spade to turn over a section of ground about six feet square for each tree. If you are doing this in the early summer, you can plant a "green manure crop", such as sorghum-sudan grass or rye, in your newly turned area to help build up the soil. A planting of buckwheat in the summer, followed by rye in the winter does an excellent job of smothering weeds while providing soil building organic matter. Again, your extension office can advise you on what grows well as green manure in your area. If it is early spring and your tree has been ordered or you have already brought it home from the nursery, some well rotted leaf mulch or compost can be added to the planting site and mixed into the soil with a shovel. If your soil is heavy and clay like, adding some coarse sand can also help drainage.

Based on your soil test results, add lime or sulphur to bring your soil pH as close to 6.8 as possible. If your soil is in need of phosphorus, add bonemeal or rock phosphate accordingly. All of these nutrients move slowly through the soil, so it is best to add them well ahead of planting. Do not add any synthetic fertilizer or fresh animal manure to the planting site if you are going to plant your tree soon. It can burn the young roots. If you have planned ahead sufficiently and are growing a green manure for a season, you can add composted animal manure or synthetic fertilizers as needed. Your soil test results will usually recommend how much you need. Prior to planting, rototill in the growing green manure so that a loose bed of soil is ready for the planting hole.

Selecting the Tree

Next, select your tree. If you have ordered from a nursery catalog, you may not be able to evaluate the quality of the tree before it arrives. When possible, order one year old trees with trunks of ½" to ¾" in diameter. (Peaches tend to run larger and often will have diameters of ¾" to 1½".) Avoid overly large trees, as they have difficulty adjusting to transplanting.

Most nurseries will send the trees "bare root", that is, with no soil or other growing medium around the roots. Reputable nurseries will take care to protect the roots from drying out, often using damp sphagnum moss or a plastic wrap. Quality trees should arrive dormant and with healthy, moist roots.

If you are selecting your tree from a local garden center, look for a tree with a straight trunk and well spaced branches with a trunk diameter as described above. Trees four to five feet in height are good. Most likely the garden center will sell trees that have been potted. Be sure that the tree is not rootbound in the pot or has many roots growing out of the bottom of the pot. This may be a sign that the tree has actually been growing in the pot for an extra season because it grew poorly in its first year.

When your trees arrive from the nursery, it is advisable to plant them as soon as conditions are favorable. If necessary, there are several ways to hold the trees for a short period. For bareroot trees, "heel in" the trees by digging a shallow trench, preferably in a shaded location. Lay the tree roots in the trench, with the trees on their sides or at an angle. Cover the roots lightly with damp soil. Balled and burlapped trees can be kept moist by wrapping the soil ball (burlap and all) in plastic. Container grown trees can just be kept in their containers, out of hot sun, and watered as needed.

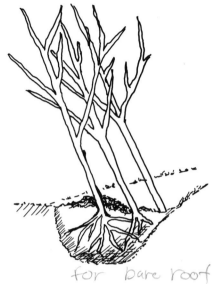

for bare root

Figure 6. *Temporarily Heeled-in Trees*

Planting   spring or fall

Fruit trees can be planted in the spring or fall, although spring planting is by far the most common. In either case, it is best to plant when the tree is dormant. If possible, plant your tree on a cool overcast day. Try to avoid planting in very wet soil. Wet soil will pack tightly around the roots and suffocate them by not allowing enough air spaces in the soil.

Start by digging your planting hole. It should be large enough to accommodate the roots without twisting them all around. A good rule of thumb is twice as wide as the rootball. Break up the sides of the hole a bit with a shovel so that the roots can grow outward later. In the center of the hole, leave a small mound of dirt on which to position the tree. At this point, an assistant is helpful for holding the tree, or you can make a handy tree support. Nail together two strips of wood at right angles, big enough to span the hole. With the support centered over the hole, tie the tree to it at the desired height. It will hold your tree until the hole is filled with dirt.

Set the tree in the hole so that it will be about two inches

Planting and Early Care 39

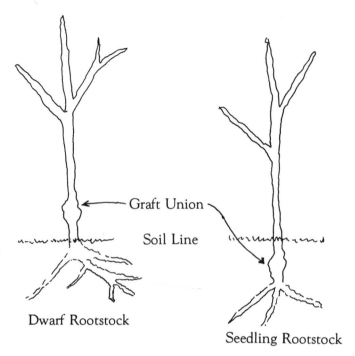

Graft Union

Soil Line

Dwarf Rootstock

Seedling Rootstock

Figure 7. Position of graft union relative to soil line

higher than it grew in the nursery. You can usually see the old soil line on the trunk. Trees that are grafted on seedling rootstocks are normally planted with the graft just below the ground. If you purchased a tree that is grafted onto dwarfing rootstock, take care to plant the graft union two to three inches above the ground level. If the graft union is planted at or below ground level, the scion variety of the tree will take root and you will loose the dwarfing characteristics. As the tree grows, the graft union will heal and appear as a bulge in the stem.

Spread the roots out uniformly around the hole. If the tree has any excessively long side roots, it is best to trim them back to fit the hole. Orient the lowest branch of the tree toward the southwest to help shield the trunk from winter sunscald. On a windy site, tip the tree into the prevailing wind three to five degrees. It will grow upright over time and be less likely to be pushed over by strong winds.

40 the Backyard Orchardist

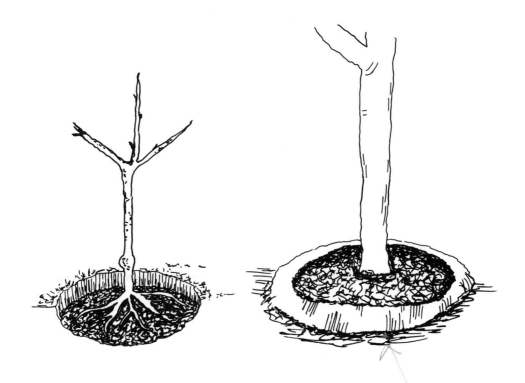

Figure 8. Tree Properly Positioned for Planting

Refill the hole with soil, gently bouncing the tree up and down a bit to fully settle the soil around the roots. Once the hole is full, firm the soil with the heel of your foot so that no large air pockets remain to dry the roots. About two feet out from the trunk, build a shallow soil dike to retain water. Mulch the dike with compost, leaves or straw. Gently water the tree until the soil is well soaked and settled. Water deeply once a week until the tree is established.

Pruning After Planting

Once the tree is planted, there are a few more little things to do to insure that it has a healthy productive future. First you will need to do some preliminary pruning.

Apple, pear, and cherry trees will be trained to a central leader or modified leader shape. To start, you will need to "whip" the tree. Remove all the side branches. With apple, pear, and sweet cherry

Planting and Early Care 41

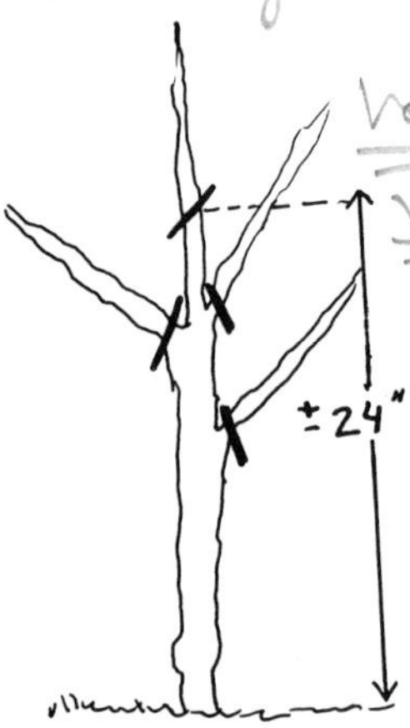

***Figure 9**. "Whipping" a newly planted tree*

only, cut the central leader (the trunk) back to about twenty four inches high. Make your cut just above a plump, healthy, vegetative bud; preferably one on the side of the prevailing wind. Do not cut the trunk of your tart cherry tree back any further. You will probably feel like you are pruning away as much or more of the tree than what you will have left. Indeed this may be the case. Most first time fruit growers are afraid that this will hurt their tree, but rest assured that this actually gives the roots a better chance to adjust to the transplanting. The new branches that begin to grow will also be stronger and better angled. When the buds begin to grow and the new branches are eight to twelve inches long, you can start selecting and training them as outlined in Chapter 11.

Peaches, nectarines and plums will be trained to the open center system. Start by cutting the trunk back to between twenty four and thirty six inches. Select three strong scaffold branches spaced about six inches apart, preferably with the top one growing into the wind. Remove all other side branches. Trim the selected scaffolds to two buds each. Chapter 11 shows how to shape your open center tree.

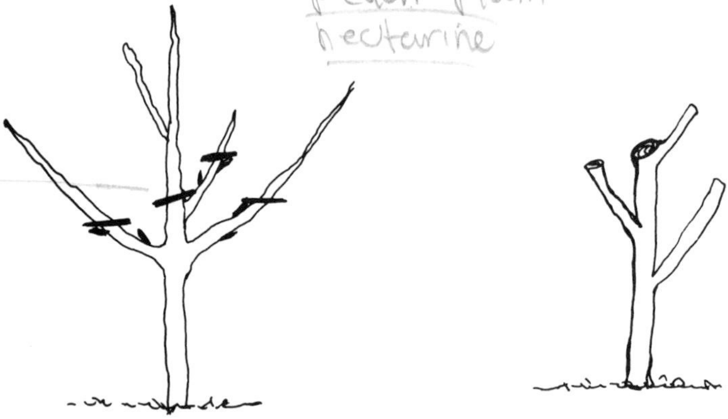

***Figure 10**. Newly planted peach pruned to an open center*

Other Care

Once the tree is planted and pruned, wrap the trunk with a flexible plastic "mouse" guard available from most nursery supply centers. Use a size that is tall enough to protect as far up to the lowest scaffold as possible. This will help protect the trunk from rodent damage for several years. Be sure that the guard does not slip below ground level in loose dirt around the trunk. Leave the mouse guard in place only until the trunk diameter fills it. Otherwise the guard will strangle the expanding tree. Usually the tree will be three to four years old before you need to remove the guard. If you have a problem with rabbits feeding in your garden you may want to replace the mouse guard with a larger, permanent rabbit guard at this point. Chapter 17 gives more information on this.

Also be sure to remove any plastic or wire labels that may have been wrapped around your tree when you bought it. As the tree grows, these will quickly strangle the trunk or branch around which they are wrapped. If you would like to keep a record of what variety your trees are, use an indelible marker and write the information on a stake that you can put in the ground near the trunk.

If you have planted a fruit tree grafted on one of the very dwarfing rootstocks, such as apple on M9 or M27, it will require a stake or trellis for support. You can install it now.

In the fall of the first growing season you should paint the trunk of the young tree with a good quality white latex housepaint. Use a brush or car wash mitt to apply paint up to the first scaffolds. On open center trees, paint the scaffolds out from the trunk nine inches too. This will protect against sunscald (also called southwest injury) in northern climates. In late winter bright sun reflects off the snow cover onto the lower trunk. The heat causes uneven expansion and contraction of the bark on the southwest side of the tree. Bark splitting often results. The white paint will reflect many of the sun's rays and keep the trunk a more uniform temperature, thus avoiding injury. Repaint the trunks every few years as needed.

Section II.

Fruit Fundamentals - Growth Habits of Specific Tree Fruit

46 the Backyard Orchardist

The Pome Fruits

In section one we discussed general characteristics of your planting location and how that would influence your fruit trees. In this section, we will discuss specific details as they pertain to each type of tree fruit. Do keep in mind what has been pointed out in preceding chapters, and think about how it relates to what you read in the chapters in this section.

We will begin by discussing the pome fruits. Although you rarely hear them referred to in this manner, you are probably actually quite familiar with them. The pome fruits are those fleshy fruits that have a central core; usually with up to ten seeds enclosed in five seed capsules. The core is protected by a thickened fleshy layer, which is usually edible, and a skin. If you are beginning to sense that we are talking about apples and pears, you are right.

Another characteristic that the pome fruits have in common is the development phases they go through each season as the buds open and develop into flowers. Very often seasonal orchard maintenance activities and insect development are correlated to these phases, so it is worthwhile to be able to identify them. The illustration on the following page should help.

Depending on the weather, the progression of these stages takes from one to three weeks. The warmer the weather, the faster these stages progress. Different varieties also develop at varying rates, but the sequence of development is the same for all pome fruit. Following fruit set, the fruit will continue to grow and develop for three to four months before it is ready to harvest.

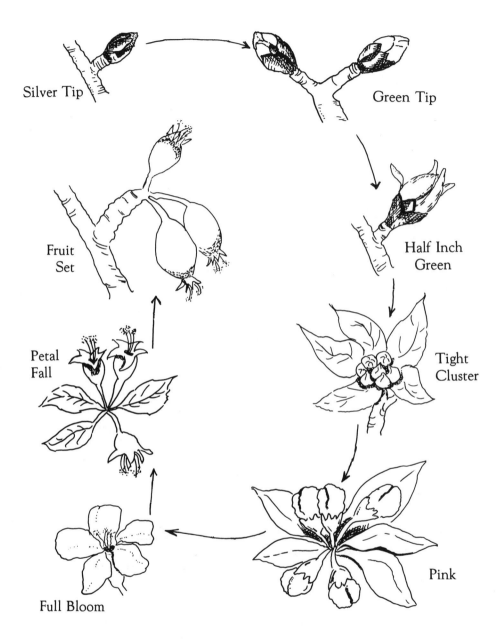

Figure 11. *Common progressive stages of development in pome fruit*

Silver Tip

Green Tip

Half Inch Green

Fruit Set

Tight Cluster

Petal Fall

Pink

Full Bloom

48 the Backyard Orchardist

4. Apples

The apple is the tree fruit most widely adapted to growing in backyard orchards throughout the country. It has many varieties to choose from and is one of the most popular eating fruits.

Most people who are unfamiliar with fruit growing will naturally assume that if you want to grow an apple tree, like Johnny Appleseed, you just plant the seed of the kind you want, let's say McIntosh, and your tree will grow. Indeed a tree will grow, but it is unlikely to be the McIntosh you expected. Like other organisms that reproduce sexually, fruit trees grown from seed most likely will not turn out to have variety characteristics identical to the tree from which they came. Due to the genetic mixing process that takes place during pollination, they can be as different from their parents as we are from ours.

Most fruit trees today are reproduced through a process called "clonal propagation". In this process many exact duplicates of a particular parent tree are made by grafting small samples of parent tissue onto a particular rootstock type. These selected rootstock pieces have also been reproduced through one of the methods of clonal propagation so that they, too, will be identical to their parent. There are several different propagation techniques, most of which the backyard gardener eventually enjoys trying out.

So, as we begin to understand the apple, we need to realize that we are actually dealing with two separate parts: the top portion of the tree or scion and the root or understock. The scion is the part that normally bears the particular variety, Delicious, Gravenstein, McIntosh, or others, that you will have to eat. It has a number of different growth characteristics that you will need to learn and work with. Likewise, the rootstock, that part of the tree that comprises the roots and lower part of the trunk, also has its own distinct characteristics. The graft union, where the scion and rootstock are joined, gives us a third area to consider.

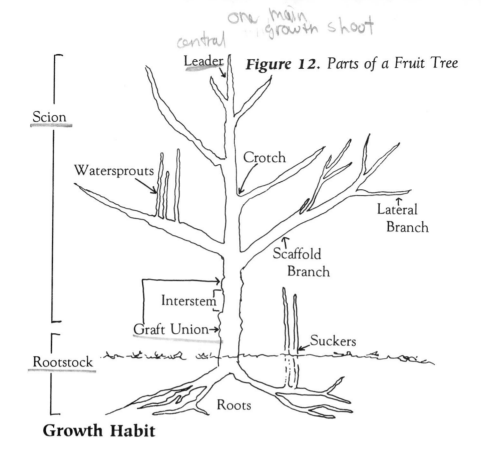

one main
central growth shoot

Leader

Figure 12. Parts of a Fruit Tree

Scion

Watersprouts

Crotch

Lateral
Branch

Scaffold
Branch

Interstem

Graft Union→

Suckers

Rootstock

Roots

Growth Habit

Let us look at the overall apple tree first. Normally the apple tree grows in a certain manner. The way and shape in which a given variety of tree typically grows is called its growth habit. For an apple tree this means that, as the tree grows up and out, it follows what is called a central leader habit. It has one main growth shoot that tends to grow straight up and become the trunk. Numerous side branches, also known as laterals, will radiate out from the trunk and form the overall "scaffold" or general skeleton of the tree. These will be the layers of horizontal branches that continue to grow out and form additional layers of branches as the tree grows taller. Within the central leader growth habit of the apple, some varieties tend to grow very upright; Red Delicious, Northern Spy, and Paulared are examples of this. Other varieties such as Cortland, McIntosh, and Empire tend to grow with more horizontally oriented laterals. Figure 13 shows some of the various growth habits you might encounter, all still within the central leader.

fig 13

50 the Backyard Orchardist

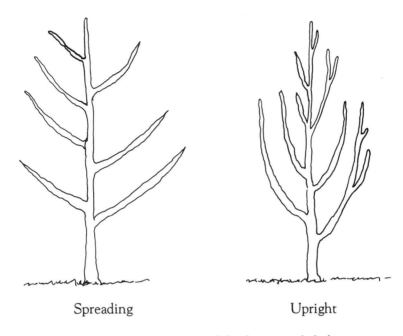

Spreading   Upright

Figure 13. *Variations of the central leader growth habit*

Flowering and Fruiting

Just as the apple tree has a certain habit of growing, it also has a certain way in which it produces its flower buds. Naturally every type of fruiting tree must produce some buds that will produce more branches (or vegetative growth) and some buds that produce blossoms (and consequently fruit). Some types of fruit trees produce all of their flowering buds on wood that was produced during the preceding growing season or, said another way, on last year's wood. Other types of trees produce flower buds on wood that is several seasons old and some produce flower buds on short, modified branches called spurs.

On apple trees most of the blossoms are borne on fruiting spurs. These spurs are generally short, compact growing branches less than four inches long. Most commonly, one year they will bear a fruit bud (or flower bud, the terms are used interchangeably) as their terminal or end bud. The following year a vegetative bud will be produced at the spur terminal and then fruiting buds and vegetative buds will alternate in subsequent years. This alternating of fruit and vegetative production will give the spur a zigzag appearance to its growth as shown in Figure 14. Of course not all spurs are producing fruit buds in the same year so there is normally an adequate supply of fruit buds to keep a crop growing every season.

fruit bud

Figure 14. Fruiting spur

Another factor to keep in mind is that, although it is not visible to you, the apple tree is already developing its fruit buds deep within its tissue the summer prior to the year in which the bloom becomes visible. Starting in about mid-June, the fruit bud tissue starts its development and differentiation. The process is completed by late March, shortly before bloom. Several factors, including the amount of light, water, and nutrients available can affect the development of the bud.

The production of flower buds alone is not enough to insure production of a successful apple crop. It is really only the first step. Once the bloom is visible, the important processes of pollination and fertilization are also going on. Pollination is the transfer of pollen from the male parts of a flower blossom to the female parts. Fertilization occurs when the male and female genetic material are united to form an embryo which can then develop into the fruit. Both of these processes are discussed in detail in Chapter 13. To complicate matters, though, a number of different fruit varieties have pollen that is not compatible with itself, the variety that is producing the pollen. This is the case with apples. Most apple varieties need to be pollinated with pollen from a different apple variety. As an example, a Red Delicious apple will not produce much fruit successfully if pollinated by another Red Delicious. This condition is known as being self-unfruitful. However, if that same Red Delicious blossom receives pollen from a McIntosh bloom, it will most likely develop a nice crop of fruit (other normal conditions cooperating). This process of sharing pollen between varieties is known as cross-pollination. Some readers may wonder what apple variety will result from this cross-pollination. The fruit variety will be determined by the blossom variety. If the blossom of a McIntosh tree has been pollinated by Red Delicious pollen, the fruit will be McIntosh. So when planting apple trees it is advisable to plant at least two varieties.

It is also important to pay attention to when the various varieties bloom, because the time that the bloom of one variety is open needs to coincide closely with the time that the pollinating variety is also blooming. Varieties that normally bloom together are found below.

Relative Bloom Time of Apple Varieties

Early	Midseason	Midseason	Late
Idared	R.I. Greening	Jonafree	Mutsu
Paulared	Priscilla	Akane	Fuji
McIntosh	Cortland	Granny Smith	Braeburn
Wealthy	Empire	Prima	Macoun
Jonamac	Winesap	Jonathan	York
Spartan	Regent	Gala	N. Spy
Liberty	Red Haralson	Honeygold	Rome
Gravenstein	Arkansas Blk.	G. Delicious	
Jerseymac	Red Delicious	Jonagold	

It is best to choose two or more varieties from within the same group, but it is possible to choose one variety from the early group and one from the middle blooming group as their bloom time will often overlap somewhat. If you choose one variety from the early group and one from the late however, their bloom times will likely be too far apart to pollinate each other well.

There is also a small group of apple varieties that, due to quirks of nature, have a triploid number of chromosomes. As a result, their pollen is sterile. Although they can be successfully pollinated by other varieties, they cannot themself serve as reliable pollen sources for other varieties. If you do want the fruit from one of the varieties listed below, be sure that your fruit planting is made up of at least three apple varieties, including those with viable pollen.

Varieties with Sterile Pollen

Arkansas Black	Red Gravenstein	Winesap
Baldwin	R.I. Greening	Zabergau Reinette
Belle de Boskoop	Sir Prize	
Bramley Seedling	Spigold	
Gravenstein	Stayman Winesap	
Jonagold	Summer Rambo	
Mutsu	Turley Winesap	

Rootstocks

As mentioned earlier in this chapter, when choosing apple trees for the home orchard one needs to consider not only the fruit variety one desires, but also the rootstock on which the tree will be growing. Of all the rootstock choices available for fruit trees, apples have the most varied array. The rootstock is such an important consideration because it will have an effect on the final size of the tree, its ability to tolerate waterlogged or excessively dry soils, its productivity and, to a lesser extent, its susceptibility to certain diseases.

Most of the apple rootstocks in use today are what is known as dwarfing rootstock. The majority of them were bred and developed at the East Malling Research Station in England from the early 1900s on. Today apple rootstocks are generally categorized into three main size groups - vigorous or standard size, semidwarf, and dwarf rootstocks. The first group, the vigorous stocks, include:

Standard - The age old "standard" seedling apple tree root. This stock is used as the measure of 100% of normal size against which the other stocks are compared. (Fruit tree size is customarily expressed as a percentage of "standard" size.) Seedling rootstocks were used in the days before dwarfing rootstocks were available and are still used today where a large winter hardy root system is desired. Since the seedling rootstock is reproduced from apple seeds rather than through clonal propagation, they may not be as uniform in their growth characteristics. The seedling root also produces a large tree, often reaching over twenty five feet tall. Consequently it is not usually the best choice for a home orchard with limited space. It is slower to produce a crop than the dwarfing rootstocks, commonly taking eight to ten years before fruiting, compared to two to five years for most dwarf apple rootstocks.

MM111 - Also in this group is the MM111 (or Malling Merton 111, named after the research stations involved in its development). MM111 is normally considered to grow to about 80% of standard size. It is drought tolerant, but also well suited for heavy clay soils that tend to be poorly drained. It has good root anchorage and wide branch angles with a very upright growth habit. MM111 is resistant to collar rot and wooly apple aphids.

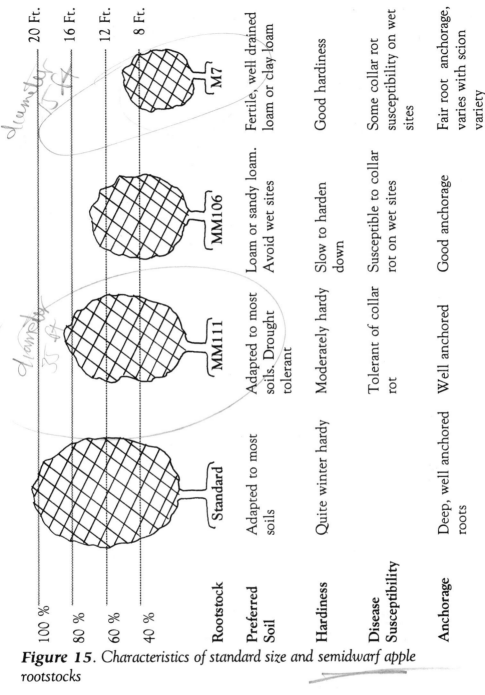

Figure 15. *Characteristics of standard size and semidwarf apple rootstocks*

The next group, the semidwarf rootstocks, are normally 50% to 70% the size of the standard apple rootstock. They include:

MM106 - Normally growing to about 70% of standard size, MM106 is one of the most popular stocks with commercial fruit growers because scions grown on it produce early in their life and bear consistently good size crops. Its strong root anchorage, size, and moderately spreading, upright shape, as well as resistance to wooly aphids, make it a very suitable combination with many of the spur type scion varieties. Several of the dwarf rootstocks have a tendency to grow many shoots or "rootsuckers" from below ground. MM106 rarely does, thus making it easy to maintain.

It is very tolerant of high temperatures and dry, light textured, sandy soils. Susceptible to the disease, collar rot, MM106 should not be planted in wet sites. Very slow to harden down in the fall, MM106 can be damaged by early winter freezes.

M7 (also known as MVIIa) - This is often considered the ideal stock for the small semidwarf tree, well suited to the home orchard. It has a nice open spreading growth habit and a good root system with strong anchorage when combined with all but a few scion varieties. At about 50% of standard size, it is similar in size to a peach tree and ideal for an espalier or decorative cordon. M7 grows well on most soils but does best on fertile loam soils and clay soils that tend to have consistent moisture. It does require regular watering on dry sandy soils. Although susceptible to wooly aphids, it is resistant to collar rot. It does have a tendency to rootsucker heavily and needs to be planted deeper than the other rootstocks to reduce this problem.

The dwarf rootstocks are the final size class of trees and offer several good choices to the orchardist with very limited space. Some of the rootstocks in this class have been available for many years, while at least one stock is a fairly new introduction.

M26 - Malling 26, at 40% of standard size, is the largest in this group. It will produce a freestanding tree on fertile soils but may need to be supported with a stake in poorer growing conditions. It produces large, early coloring fruit. Susceptible to wooly aphids and fireblight, it also does not tolerate wet soil conditions. Some

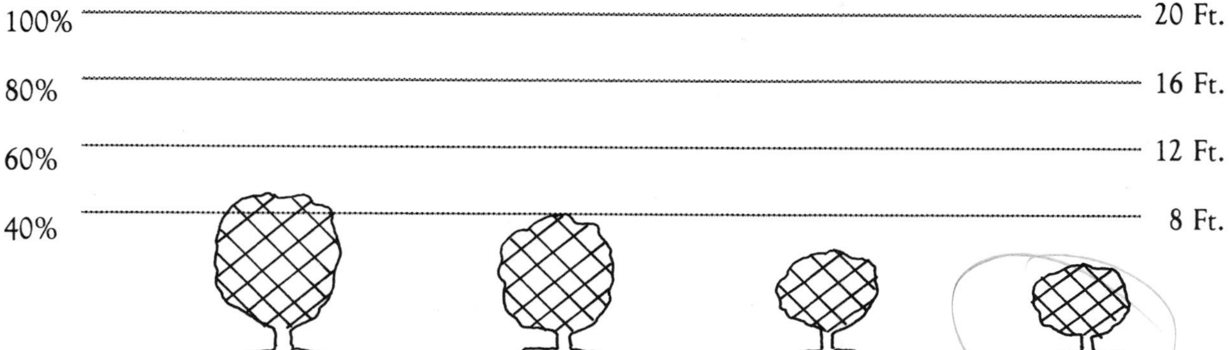

Rootstock	M26	Mark	M9	M27
Preferred Soil	Fertile loam Avoid wet site	Prefers heavier soils	Well drained fertile soil	Fertile loam
Hardiness	Good hardiness	Moderately hardy	Moderate to good hardiness	Unknown, still testing
Disease Susceptibility	Very fireblight susceptible. Slightly collar rot susceptible	Collar rot resistant	Collar rot resistant	Fireblight resistant
Anchorage	Freestanding but benefits from support	Varies with site	Needs trellis or stake for support	Needs trellis or stake for support

100% — 20 Ft.
80% — 16 Ft.
60% — 12 Ft.
40% — 8 Ft.

Figure 16. *Characteristics of dwarf apple rootstocks*

graft union breakage has been experienced with a limited number of scion varieties. So, it is wise to inquire before purchasing trees with this stock, although most reputable nurseries avoid propagating these problem combinations. In spite of its imperfections, M26 is a very good rootstock for the backyard orchardist with limited space. It exhibits good winter hardiness and better root anchorage than one of its parents, the more dwarfing M9 rootstock. It can start bearing a crop when it is three to four years old.

Mark - (tested as Mac® 9 from the Michigan Apple Clone series) Mark, the newest introduction to the dwarf rootstocks, is not widely available to the home orchardist yet. As it gains exposure in the commercial fruit industry, more wide-spread testing is showing its attributes to vary greatly with site and management. Use Mark on a limited basis at this time, budded to a vigorous scion, and only if you can provide a moist site and intense management. Slightly smaller in size than M26, Mark offers the benefit of early cropping but requires thinning for adequate fruit size. Its small root system requires careful attention to watering and trunk support. Mark is less susceptible collar rot than M26.

M9 - Most dwarfing of the readily available rootstocks, M9 grows to about 20% to 30% of standard size. It bears fruit very early in its life (often within one to two years of planting) and in abundance. It requires a good fertile site and careful watering as it has a rather small, brittle root system. Resistant to collar rot, M9 prefers cooler soils and benefits from mulching. M9 needs to be staked or supported by some type of trellis. It is very well suited to a decorative espalier.

M27 - Relatively new and most dwarfing of the Malling rootstock series, M27 grows to a maximum height of six feet. It requires staking or trellising for support. It is resistant to fireblight, but not fully tested yet for cold hardiness. It is also being tested in combination with MM111 as an interstem to produce a larger freestanding tree.

In addition to the extensive rootstock selections that have come from the East Malling breeding program, there are several rootstocks that have come from programs in Canada, Poland, Russia, and Sweden. All of them are particularly noted for their cold hardiness. They include:

Alnarp 2 - From the Swedish breeding program, this rootstock produces a large tree similar in size to apple seedling. It is drought tolerant and has excellent root anchorage.

Budagovski 9 - A Russian selection, Bud. 9 is used mostly as an interstem producing a tree about 40% of standard size. Although very hardy, its roots are small and brittle, hence its use as an interstem. Used as a rootstock, staking is recommended.

Ottawa 3 - This rootstock produces a tree slightly smaller than M26, but better anchored. It is also more resistant to dogwood borers and winter cold injury. Fruit on this fairly new selection from Canada matures two to three days earlier than on most other rootstocks. Unfortunately, it is difficult to propagate making its current supply limited and sometimes difficult to find.

P-22 - A new selection from Poland, P-22 produces a tree similar to M27 in many ways. Small in size, P-22 has very brittle roots and graft unions. Reports indicate it can survive temperatures as low as -40° F.

All of these very hardy rootstocks have only recently found their way into the marketplace. Other crosses from these breeding programs may eventually be released too. For northern climates, they are certainly worth considering on an experimental basis. As supplies become more plentiful, they will also be easier to locate.

Variety Choices

In selecting from the rootstocks we have just discussed, your choices may be dictated and limited by what conditions exist on your particular site. In choosing the scion variety however, you have many to choose from.

To minimize the amount of care and expense required and reduce pesticide exposure, strongly consider planting the new disease resistant varieties. For backyard fruit growers, the majority of spraying

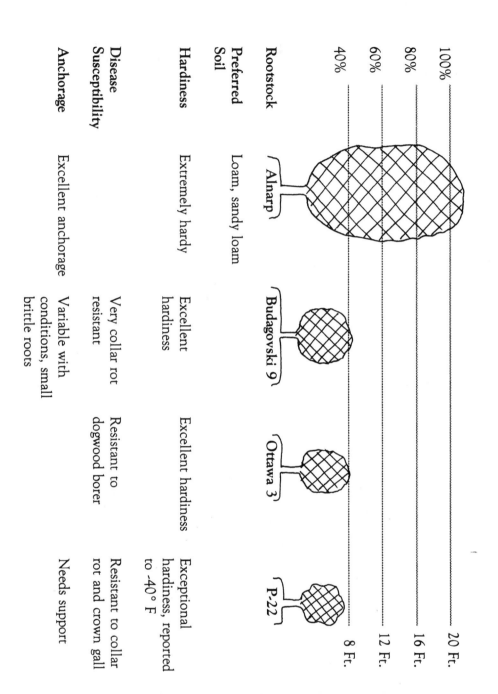

Figure 17. *Characteristics of hardy apple rootstocks*

that is done is aimed at controlling fungus diseases. A number of varieties being sold are resistant to one or more of the major apple diseases. By choosing varieties that are resistant to the major disease problem in your area, you may easily be able to reduce your seasonal number of sprays from six or eight to only one or two.

Disease Resistant Varieties

Variety	Scab	Cedar/Apple Rust	Mildew	Fireblight
Akane	R	MS	R	VS
Baldwin		VR	VS	
Beacon		R		
Dayton	HR	S	R	R
Enterprise	HR	R	R	R
Freedom	HR	R	R	HR
Gold Rush	HR	S	R	MR
Grimes Golden	R	VS	S	R
Jonafree	HR	S	R	R
Jonathan	MR	MS	VS	VS
Liberty	HR	HR	R	HR
Macfree	HR	MS	R	R
N. Spy	S	R		R
Nova Easygro	HR	R	R	R
Novamac	HR	R	R	MS
Prima	HR	VS	R	VS
Priscilla	HR	HR		HR
Redfree	HR	HR	R	R
Sir Prize	HR	S	S	S
Spartan	R	MR	MR	MR
Sweet Sixteen	R	S		S
Williams Pride	HR	R	R	R

HR=Highly resistant, R=Resistant, MS= Moderately resistant, MS=Moderately susceptible, S=Susceptible, VS=Very susceptible

Black rot, bitter rot, sooty blotch, and fly speck can be a problem in humid areas. Akane, Arkansas Black, Blairmont, Fuji, Gala, Melrose, and Stayman Winesap are fairly resistant to these "summer diseases".

That you will need to pay attention to cold hardiness and pollination requirements has already been pointed out. Another factor that you will need to keep in mind is whether the growing season in your area is long enough to properly mature a particular variety. A look at Figure 4 will give you the number of frost free days commonly found in the growing season in your area. Your best bet for a reliable, tasty crop of apples (as well as the other fruits in this book) is to choose varieties that require close to this many growing days. Occasionally you may have a longer season than indicated on the map. However, if you choose a variety that normally requires a longer season to mature, you will have apples to pick, although they will not be all that flavorful and, in the unusually short season, may not even be ripe enough to pick before hard frosts hit. Also be aware that certain types of apples favor certain growing climates. McIntosh and its related types (Macoun, Jerseymac, Liberty, Paulared) like warm fall days and cool nights for development of red color. They grow best in New England and upper New York. Jonathan types (Jonalicious, Jonamac, Jonagold) favor warm temperatures following bloom and grow well in the Central United States (Missouri, Ohio, Illinois).

If you are planting in the confines of a small yard, consider varieties with a spur type growth habit that is more compact than the lanky growing non-spur varieties. They will require less space and are often easier to prune and harvest too. With all of these considerations said and done, it's time for what many backyard growers relish -- actually choosing the varieties. Your choice will depend on what you intend to do with your harvest: eat it naturally, crisp, crunchy and sweet straight from the tree; perhaps cook with it, preserve it, or make cider with it. With so many apple varieties to choose from, it is not surprising that some will be better suited to certain uses than others. The charts that follow outline what are often considered to be some of the best flavored varieties and their best uses. The varieties listed are all available through one or more of the sources listed in the appendix. Since there are so many fruit varieties, other choices are also available, so don't limit yourself if you find one elsewhere that interests you and is suited to your area. (Likewise, in the interest of serving readers of future editions of this book we would love to hear of your discoveries and include them in future updates. Please send us a note to bring them to our attention.)

Popular Apple Varieties and Their Best Uses

Variety	Eating	Pie	Baking	Drying	Sauce
Arlet (Swiss Gourmet)	E				
Blushing Golden	E *				
Braeburn	E	✓	✓		
Calville d'Hiver	E	E	E		
Cortland	E	E	E		C
Empire	E	E	E		C
Fiji	E				
Gala	E		✓		✓
Ginger Gold	E				
Gold Rush	E				
Golden Delicious	E	✓			
Gravenstein	E	✓	✓		C
Haralson	✓	✓			✓
Holstein	E	E	✓		
Idared	✓ *	E	E		✓
Jonathan & types	E	E	✓		C
Jonagold	E	✓	✓	✓	
Jonalicious	E				
Jonamac	E	✓	✓		C
McIntosh & types	✓				S
Jerseymac	✓	✓			S
Liberty	✓		✓		S
Macoun	E		E		
Spartan	E	✓	✓		S
Melrose	E *	E	E		✓
Mutsu	E	✓	✓		✓
Northern Spy	✓	E			
Red Astrachan	✓	E			
Red Rome	✓ *	✓	E	✓	
Rhode Island Greening	✓	E	E	E	
Senshu	E				
Spigold	E		✓		✓
Summer Rambo	✓	✓		✓	✓
Wealthy	✓	✓			✓
Yellow Transparent		✓		✓	S
Yellow Newtown	E	✓			

E=Excellent for use, *=Best after storage, ✓=Good for use

Just as apple varieties provide a range of tastes from sweet to tart, they also provide a range of textures, especially when cooked. Some are firm and crisp, others soften quickly after harvest. Most of the preceding varieties indicated as being suited for pies tend to maintain a strong apple flavor and fairly firm slice after cooking. The sauce apples can be divided into two classes - smooth and chunky. They are coded with a "C" or "S" to indicate the type of sauce they produce. In general, the early season fruit (those ripening before McIntosh) will produce a smooth sauce. Late season varieties will be chunkier. Also, many experienced cooks have discovered that the secret to a superb sauce is in blending varieties of several flavors and textures, much the way a blended cider is produced.

Pressing cider is an excellent way to make use of surplus or blemished fruit, but it is also an art. Indeed cider connoisseurs would argue that it warrants its own chapter rather than a mere section. Much of what gives a really good apple cider its magic flavor comes from the fact that it is a subtle blend of several factors that effect our taste buds: sweetness, tartness, astringency and aroma. Most of the best ciders will combine several apple varieties that bring together these factors. Through experimenting you will come up with your own "secret recipe" that you enjoy most. Try starting with a mix that is two parts sweet, two parts tart, one part aromatic, and one part astringent varieties. Appreciate, too, that each growing season is unique and will influence the fruit somewhat differently from year to year. This helps give each new batch of cider its own unique flavor. Here are a few possibilities to get you started. Drink up and enjoy!

Cider Apple Varieties

Tart	Sweet	Aromatic	Astringent
Cortland	Gold Delicious	Jonathan	R. I. Greening
McIntosh	Empire	Grimes Golden	Fameuse
Idared	Red Delicious		Russet
Spartan			Crabapples
Spy			

Varieties for Special Situations

Although apples are the fruit most easily grown throughout the United States, certain locations still have climatic conditions that call for specially adapted varieties. Hardy apple varieties that are particularly able to stand low winter temperatures and still produce a reasonable crop are listed below. Coupled with one of the cold hardy rootstocks this should enable you to grow apples in such areas as northern Minnesota and the upper Plains and Mountain states.

Hardy Apple Varieties

Anoka	Fameuse	Keepsake	Regent
Beacon	Fireside	Macoun	Sweet Sixteen
Connell Red	Gldn. Russet	McIntosh	Wealthy
Cortland	Haralson	Norland	
Davey	Honeygold	Northern Spy	
Earliblaze		R. Astrachan	

Warm winter climates, such as southern California and Florida, require varieties that have low chilling requirements. These varieties will satisfy their dormant rest needs with less than 600 hours below 45° F. Some of most popular varieties are listed below.

Low Chill Apple Varieties

Anna(400)	Ein Shemer(400)	Wealthy
Beverly Hills(400)	Gordon(350)	Winter Banana
Braeburn	Granny Smith	White Winter
Dorsett Gold(100)	Lady Williams	Permain(400)

One last factor you may encounter as you look around is that most of the retail nursery catalogs generally sell varieties that have a broad appeal across the country. In searching out local sources, you may stumble upon a locally popular old fashioned (sometimes called antique) variety that is well adapted to your area. Even though it may not be listed in this or any other book, certainly don't be afraid to try it. That is one of the joys the home orchardist has -- you can experiment with the unusual and often reap special pleasure from it!

Average years to bearing: Standard, 8 years
Dwarf, 3 to 6 years

Average yield per tree: Standard, 10 to 20 bushels
Dwarf, 1 to 6 bushels

Space needed per tree: Standard, 25 to 30 foot circle
Dwarf, 6 to 15 foot circle

Average mature height: Standard, 25+ feet
Dwarf, 6 to 20 feet

Days from bloom to harvest: 90 to 180 days

Pollination requirements: Cross pollinate with another apple

Most common pruning system: Central leader

Commonly used rootstock: Seedling root for standard size tree
M7 most versatile for home use

Common insect pests: Curculio, codling moth, apple maggot

Common diseases: Scab, mildew, cedar apple rust

Disease resistant varieties: See chart page 61

Popular varieties & best use: See chart page 63

Useful life: Standard, 50+ years
Dwarf, 20+ years

Suggested number of trees for family of four: 2 - 3 trees

Bears fruit on: Spurs that are several years old

66 the Backyard Orchardist

5. Pears

Like the apple, the pear is a pome fruit with a fleshy body surrounding a core of seeds. Many of the principles of growing apples will also apply to pears. Normally a pear tree may take as much as eight to ten years before bearing a substantial crop, but the tree can live and produce reasonably for over 100 years. In general, pear trees require less care than apple trees and are therefore an excellent choice for the home orchard.

Similar to the apple, the pear has a central leader growth habit. It tends to grow more upright than many apple varieties. Fruit is born on spurs, similar to the apple. Since upright shoots do not readily produce lateral growth, they need to be encouraged to bear spurs and side branches by training them to a more horizontal position. In many home gardens, these flexible new shoots can be trained to a trellis or espalier to form an attractive hedge row or specimen tree. With careful hand thinning of the fruit, a crop that is both beautiful to look at and wonderful to eat can be achieved fairly easily. Due to this, the pear was often the fruit of choice in courtyards and gardens of European nobility of the 18th and 19th century.

If you choose to grow your pear as a free standing tree, it is normally recommended that it be trained to a modified central leader when it is young. Six or seven lateral scaffold branches should be encouraged to develop and form the main structure of the tree. When the tree has reached a manageable height (say around ten feet) the leader is headed off to a strong lateral branch.

Since many pear varieties are quite susceptible to the bacterial disease fireblight, it is best to prune the young tree sparingly. The pear tree tends to be naturally quite vigorous and will often grow more than twenty four inches in one season. The vigorous shoots tend to be very upright in their growth and the tissue quite green and succulent. This makes the tissue particularly attractive to the fireblight bacteria and the sucking insects that spread them. To avoid

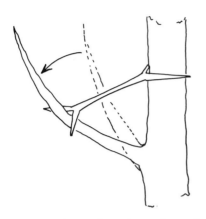

problems with fireblight, it is best to minimize heading back cuts that encourage profuse lateral growth. It is more desirable to encourage side branching by spreading the scaffolds with wooden or plastic spreaders or tying them down with twine. Light annual thinning out cuts can be used to eliminate unwanted shoots. Once the pear tree starts producing, the weight of the fruit will also help spread the branches.

Figure 18. *Spreading branch with a plastic spreader*

Care should be taken not to encourage very lush growth in pear trees, especially during the early spring and during bloom time when ideal disease development conditions often exist. Feeding the pear tree only small doses of nitrogen and light annual dormant pruning are two of the easiest ways to avoid fireblight.

Pollination

Like apples, most pears tend to be self-unfruitful and require pollination by a variety other than themselves. Even the self-fruitful varieties produce a larger crop when cross pollinated. For some reason, pears have a flower that is not particularly attractive to bees (the main insect active in pollenizing). Therefore it is often recommended that two other pollen varieties be available in pear plantings. Since a number of pear varieties are so distinct from each other, it is actually quite nice to include three different ones in the home orchard and enjoy their harvest at different times during the season.

In choosing pear varieties as pollen sources, keep in mind that Magness has sterile pollen and therefore will not function as a pollinator. Also, Bartlett and Seckel will not cross pollinate each other. Bartlett is sometimes considered to be partially self-fruitful and if you can really only fit one pear tree in your yard, you may want to make it a Bartlett unless fireblight is a serious problem in your area. Honeysweet and Moonglow are two varieties known for their strong pollen.

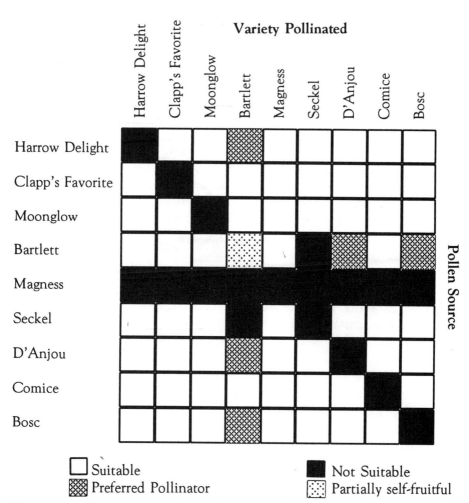

Figure 19. Pear varieties that pollinate each other well

Rootstocks

The number of pear rootstocks in use today is not as varied as the apple rootstocks, but the reasons for choosing the rootstock will still be similar. In the home orchard you will most likely want the size controlling characteristic of a dwarfing rootstock. Disease resistance will also be important. The major pear rootstocks available today are:

Pyrus calleryana - This rootstock will grow the largest pear trees at about 115% of standard seedling size. Its strongest attributes are that it is quite tolerant of heavy, wet soils and resistant to fireblight, wooly aphids, and nematodes. In locations with cold winters, it is not as winter hardy as the other stocks and is best suited for use in the southern United States and the west coast.

Bartlett Seedling - Bartlett seedling rootstock is normally considered to be the standard 100% size that other pear stocks are measured against. The most commonly used pear rootstock, Bartlett seedling is vigorous growing and winter hardy. It has a well anchored root system, forms excellent graft unions with all of today's commercially available varieties and is well adapted to wet soils. Its main fault is it's high susceptibility to fireblight.

OHxF-333 (OHxF-513) - These rootstocks are numbered selections from crosses of the Old Home and Farmingdale pear varieties. At about 70% of standard size, they bear early, are fireblight resistant and quite winter hardy. OHxF-333 performs quite well as a rootstock for Asian pear varieties.

Quince - As a series, quince rootstocks are among the most dwarfing available for pear today. Quince A is abut 50% of standard size and Quince C produces a tree as small as 30% of standard. They tend to be susceptible to freeze damage in northern areas but are suited for use on the west coast and southern east coast. Not all scion varieties can be compatibly grafted to quince. To overcome graft union incompatibility with Bartlett, Bosc, Seckel, and Clapp's Favorite; an interstem of Old Home is often used. Although resistant to nematodes, Quince rootstocks are not well adapted to heavy wet soils.

Domestic Pear Varieties

When most people think of pears and what to do with them, they think primarily of eating them fresh or canned. However, pears are excellent for other uses too. The adventurous home orchardist would do well to dry some of the harvest for use in cooking or just plain snacking. Baked pears, marinated in wine are an elegant and simple treat too. Another nice thing about pears is that, unlike a number of the earlier summer fruits, they ripen over a fairly extended time period. Different pear varieties are better suited to certain uses than others. In order of ripening they are:

Harrow Delight - Ripening in early to mid August, Harrow Delight is a medium sized, high quality eating pear. It is resistant to fireblight.

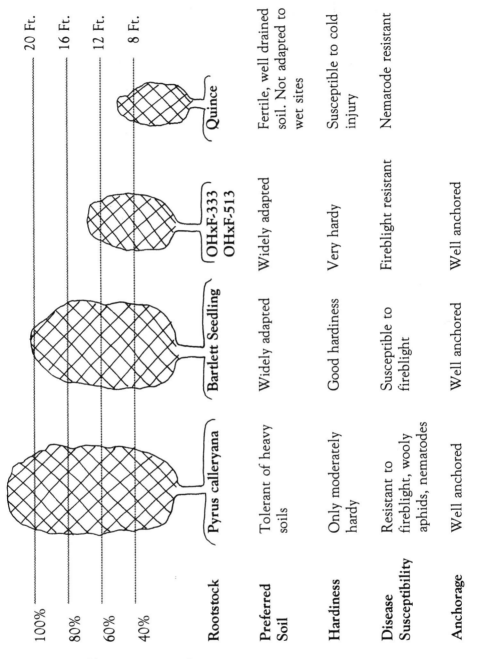

Figure 20. *Characteristics of pear rootstocks*

Chapin - One of the best flavored early pears, Chapin has a smooth, sweet, melting flavor. A productive, compact tree, it does well in the home orchard.

Clapp's Favorite - A large, attractive, juicy pear, Clapp's Favorite is quite productive and very hardy. The tree has a more open spreading growth habit than many other pear varieties and produces an abundance of spur growth. The tree is quite susceptible to fireblight. It should be pruned sparingly and during the dormant season only. Fruit ripens in mid to late August and has a tendency to have internal breakdown at the core, so it is important to pick Clapp's Favorite early. It is excellent for eating and acceptable for canning. Because of its sweetness, you may be able to use a lighter syrup when canning this variety. It's texture, when canned will sometimes be grainy and not quite as smooth as Bartlett.

Moonglow - A medium sized, mild flavored pear, Moonglow is tolerant to fireblight and is an upright vigorous grower. Production is increased by planting near a pollinator variety. Moonglow ripens about a week before Bartlett and bears early in its life.

Aurora - An extremely high quality eating pear. Large in size, Aurora will keep for several months. The tree is susceptible to fireblight with a vigorous spreading growth habit.

Bartlett - The world's best know all-purpose pear variety, Bartlett is sometimes called "Williams" in Europe and in older literature. It has the stereotypical pear shape and a nice juicy, sweet flavor. Its smooth texture has made it a favorite for canning both in the home kitchen and for commercial production. The tree is adaptable to a number of climates, but is susceptible to fireblight. The ripening season for Bartlett varies from early August in California to early September in Michigan and much of the northeast. It does hold better on the tree than most other pear varieties, but still keeps best if picked green and ripened off the tree.

Magness - A fireblight tolerant pear, Magness has medium sized fruit with a slightly russetted skin that is fairly tough and resistant to insect injury. Normally harvested in early September, the fruit will store well for up to three months and still ripen to acceptable quality. Magness will start producing at around six to seven years of age and will require a good pollen source as it is pollen sterile. Magness can be planted with many of the Asian pear varieties since many of them will pollinate Magness.

Seckel - A very hardy and productive self-fruitful pear, Seckel has long been a favorite of home gardeners. Often called "the sugar pear", it is small, firm and very sweet. It is excellent for fresh eating and very nice for making spiced or pickled pears. A bit slow to begin fruiting, it is somewhat resistant to fireblight. The normal harvest time for Seckel is early to mid September.

D'Anjou - This green, large sized pear has excellent quality as a winter eating pear and will store as long as seven months under proper conditions. A large vigorous tree, D'Anjou is winter hardy and fireblight tolerant. Due to its high vigor, it will benefit from heavy, selective pruning to maintain fruiting spurs. D'Anjou has long been a favorite for the home orchard.

Comice - An excellent flavored eating pear, Comice is not normally used for canning. It is selective in where it grows best, favoring Oregon and California primarily. Normally harvested in late September in these areas, it has a tender skin that requires careful handling. Comice is vigorous and fireblight resistant, but tends to crop erratically.

Beurre Bosc - A premium quality eating pear, Bosc has a heavily russetted skin and very smooth white flesh. Normally harvested in early October, it keeps well and will develop considerable flavor in storage. A slow growing tree with a leggy growth habit, it is sometimes difficult to train. The tree will be large and productive at maturity. Somewhat susceptible to fireblight, older Bosc trees are also occasionally more prone to stony pit virus.

Cold Hardy Pear Varieties

Beurre Giffard	Clapp's Favorite	Lincoln
Beurre Hardy	Golden Spice	

Fireblight Resistant Pear Varieties

Ayres	Lincoln	Old Home
Douglas	Magness	Orient
Duchess	Maxine	Seckel
Golden Spice	Mericourt	Tyson
Harrow Delight	Moonglow	Ure
Kieffer	Morgan	Warren

Judging Ripeness and Harvest Time

A special note about harvesting pears should be made. While most fruits are at their most flavorful if allowed to remain on the tree until fully ripe, domestic pears are an exception. A pear ripens from the core out, so that although it may appear hard and green on the outside, the inside is often already overripe, broken down and has developed a musty unpleasant flavor. How is it then that you may have eaten some very sweet delicious pears?

If pears are picked when they are still hard and slightly immature and then partially ripened under refrigeration, you will get a fruit that has both excellent flavor and keeping quality. Observe your pears daily as they approach the normal ripening time in your area. (If this is your first pear crop, you may want to consult your local extension service regarding normal harvest dates in your area.) The number of days from bloom to harvest is also a reliable indicator of harvest readiness. The chart below lists some of the common varieties and their normal maturity time.

Days from Harvest to Bloom - Pears

Variety	Days from bloom to harvest
Anjou	140-165
Bartlett	110-135
Bosc	150-165
Clapp's Favorite	105-130
Comice	150-170
Kieffer	170-190
Seckel	120-140
Winter Nelis	160-185

Just as the ground color of the skin begins to change from a dull green to greenish-yellow the fruit should be ready to pick. Most often you will notice this change first around the lenticels, the small pores in the skin. When the color of the lenticels changes from white to brown the pear is mature and will ripen off the tree. Pick the pear with a slight upward twist to one side. If the fruit is ready to be picked, it will pop off the spur fairly easily. Some of the early season varieties may need to be harvested in several pickings over the course of a week's time. Varieties from Bartlett season on can usually be harvested in one picking. Once the fruit is picked, store it at 34° F for two to three weeks and ripen small quantities at room temperature to suit your needs. To check for eating readiness gently press around the base of the stem with your thumb. If the flesh gives slightly, the pear is ready to eat. Don't allow it to become overripe. Also do not use yellow color as an indicator of ripeness because the fruit will often be overripe and brown inside before the outside is very yellow. Pears that are not fully ripe can be stored anywhere from several weeks to several months if kept properly cool. Ripe pears can still be kept several days if refrigerated, but flavor may decrease some.

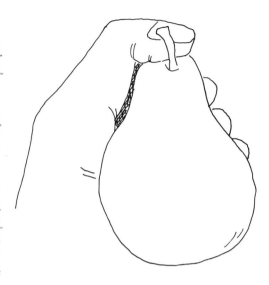

Figure 21. Testing for ripeness

Asian Pears

A whole new category of pears that is becoming very popular in the gourmet market is the Asian pear, sometimes also called the pear-apple, salad pear or sand pear. As easy to grow as the domestic types, the Asian pear's unusual appearance sets it off from the domestic pear in that most varieties are somewhat smaller and fairly round in shape. The flesh is firm and crunchy, more like an apple, and the flavor is extremely sweet and juicy. Where the domestic pear has a smooth buttery flavor, the Asian pears usually explode with a juicy

Figure 22. Asian pear

Pears 75

sweetness. They have a long storage life and most varieties easily keep for four to five months under proper conditions.

Asian pears have several other characteristics that set them off from the domestic pears. They tend to be fairly precocious, beginning to bear when they are often only two to three years old. Their fruit set is often very heavy, which can lead to large crops of small fruit if not thinned diligently. Due to this heavy cropping, they can easily develop a biennial bearing habit of heavy crop one year, no crop the next. The best cure for this is to remove many of the newly forming fruit shortly after bloom. Fruit should be thinned to one fruit in each cluster, preferably spaced five inches apart. Leave a maximum of fifty fruit on a young (three to four year old) tree increasing that gradually to about 250 fruit per tree by the age of seven. A mature Asian pear tree can bear as much as 400 pounds of fruit in a season or close to 800 pears weighing a half pound each. Production may continue for anywhere from 50 to 100 years.

Along with the pleasure of their fruit, the Asian pears can also be an attractive addition to the fall landscape. The foliage of many cultivars turns a deep purplish red and can be a wonderful autumn focal point in the yard.

As a group, the Asian pears are usually subdivided into the Japanese varieties and the Chinese varieties. The skin of Japanese pear varieties is usually either russet free with a green undercolor that turns yellow with maturity or russetted with green or yellowish brown skin. So you will often see them classified as either russet-free or russet-skinned varieties. Depending on the variety, the Japanese pears ripen anywhere from late June to mid October in North America. Most of the Asian pear varieties available from nurseries in the United States are of the Japanese type.

The Chinese pear varieties tend to have more of the traditional pear shape and most have smooth green skin. Their flavor also tends to be milder than the Japanese strains. Chinese varieties may be more suited to areas of the south, where the winter is short and mild, as they have a very short chilling requirement to break dormancy. Consequently, they have a tendency to bloom early in the spring. Although they are winter hardy, their early bloom could be a problem in northern areas with late freezes. In the south, though, they still bloom after peaches, and usually escape most of the frosts. A few of the more commonly available varieties today are:

Hosui - This variety has russet skinned golden brown fruit with a mild sweet and juicy flavor that has rated very highly in taste tests. Late blooming, Hosui is self-fruitful, but will still benefit from cross pollination. It starts bearing fruit fairly young and is fireblight susceptible.

Niitaka - A moderately fireblight resistant variety, Niitaka produces fruit that have a relatively short storage life. It's flowers have sterile pollen and therefore, it is not suited as a pollinator.

Shinko - A late harvested variety, Shinko has excellent storage life of up to 6 months. It begins bearing early and with heavy crops that require diligent thinning for well sized fruit. It is the most fireblight resistant of the Asian pears and also resists blossom blight relatively well. The tree is relatively small and upright. If you have space for only one Asian pear, Shinko may be the best choice.

Shinseiki - A crisp textured, sweet, juicy, white fleshed pear, Shinseiki is round and yellow with very little russet. An early ripening variety, normally picked in mid-August, the tree is medium sized and spreading. It needs another Asian pear or Bartlett as a pollinator.

Twentieth Century - A mid-season pear, also known as Niji-seiki, it is normally harvested in late August and September. The fruit is firm and juicy with a mild, slightly tart flavor. Twentieth Century has a lopsided round shaped fruit of uniform size. Early reports show it to be quite productive and in need of fruit thinning. Its profuse, showy bloom make it a wonderful land-scape accent. It is somewhat fireblight susceptible.

YaLi - This is reputed to be the best tasting of the Chinese pears. It does best in warm climate areas, blooming early in the spring and having a chilling requirement of only 300 hours. It is fireblight tolerant, but slow to start producing fruit.

Care of Asian Pears

Care of the Asian pears is similar to domestic varieties. On the whole, Japanese varieties seem to be as susceptible to fireblight and pear psylla as the domestics. For fireblight resistance, consider planting the Chinese strains or Japanese-Chinese crosses. Asian pears are also susceptible to pseudomonas blossom blight. This can be a problem where cool wet weather is prevalent during bloom, as is often found in the Pacific northwest states. Preventive treatment for both of these major diseases should be considered depending on your local conditions.

Most Asian pears require cross pollination. Either another Asian pear or a domestic variety is suitable as a pollen source as long as their bloom periods coincide. They are quite productive, and hand thinning of excess fruit is advised.

As a group, Asian pears are quite vigorous and should be given nutrients (especially nitrogen) and water sparingly to avoid too much lush, succulent growth. Some varieties are prone to iron or magnesium deficiency. Zinc deficiency can be a problem with Asian pears in the western states. Symptoms are similar to those indicated in Chapter 10.

Rootstocks used for Asian pears are for the most part the same as those used with the domestic varieties. Early reports show the scion wood to be quite winter hardy, so the rootstock's hardiness and soil requirements will likely be the limiting factor in selecting a combination for your site.

Harvesting Asian Pears

Harvest timing and care is probably one of the areas where Asian and domestic pears differ the most. Unlike domestic pears, Asian pears should be ripened on the tree. Watch for a gentle color change from green to light brown or green to yellowish green as a signal that the fruit is approaching ripeness. Also taste test every few days and pick when flavor is sweet. Since most Asian pear varieties ripen unevenly, you will probably have to selectively harvest ripe fruit over several pickings. Handle fruit very carefully as Asian pears have a very tender skin that shows bruises very easily. Brown discolored marks will show up if the fruit has been roughly handled. Bruises also decay more easily. When properly harvested, the fruit can be kept for two to three weeks at room temperature and up to six months, depending on variety, under refrigeration. If fruit turns "punky" or spongy in storage, it was probably underripe when harvested.

Since many of the Asian strains have a fairly tough skin, they are often eaten peeled. Many are also sliced away from the core to avoid the gritty stone cells found there.

QUICK REFERENCE
Pears

Average years to bearing: Standard, 8 to 10 years
                             Dwarf, 3 to 5 years

Average yield per tree: Standard, 3 Bushels
                             Dwarf, 1+ Bushel

Space needed per tree: 10 to 15 foot circle

Average mature height: 15 to 20 feet

Days from bloom to harvest: 120 to 190 days depending on variety

Pollination requirements: Need cross pollination

Most common pruning system: Modified central leader

Commonly used rootstock: Bartlett pear seedling

Common insect pests: Psylla, codling moth, curculio; stink
                             bugs on Asian pear

Common diseases: Fireblight, pear scab

Disease resistant varieties: See chart page 73

Popular varieties & best use: Bartlett - canning, eating
                              Bosc & D'Anjou - storage, eating

Useful life: 15+ years

Number of trees for a family of four: 2 trees

Bears fruit on spurs that are several years old

Stone Fruit

The following chapters will discuss some of our favorite dessert fruits: cherries, apricots, peaches, nectarines, and plums. There is an expression often heard in Traverse City, Michigan, The Cherry Capital of the World: "Life without cherries is the pits". So it is and so we also find out what all of these prized fruit have in common. They all have a single pit or stone inside as compared to the core of seeds found in the pome fruits just discussed.

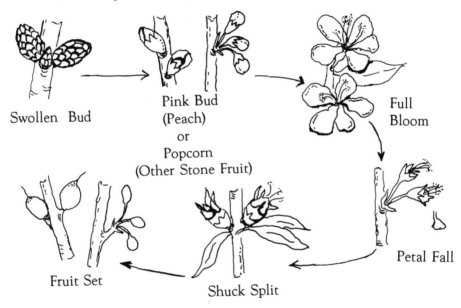

Swollen Bud

Pink Bud (Peach) or **Popcorn (Other Stone Fruit)**

Full Bloom

Petal Fall

Shuck Split

Fruit Set

Figure 23. Common growth stages of stone fruit

Just as the pome fruits shared several common developmental stages, the stone fruits as a group also share a corresponding series of seasonal growth phases. Figure 23 will help you identify them.

Most stone fruit tend to bloom early in the spring and are therefore challenging to grow in some locations. Although most stone fruit trees are relatively winter hardy once properly acclimated, many of them are susceptible to winter cold injury for two major reasons. One, most stone fruits have a relatively brief chilling requirement and may start to grow again during an unexpected mid-winter warm spell. Sweet cherries and apricots are often subject to this problem. Peaches in particular can suffer from winter injury of a second type. Since they grow so vigorously throughout the season, peaches often still have tender green tissue early in the fall. An early cold snap can be quite damaging if this tissue has not had time to acclimate slowly and "harden down". Comments in the following chapters regarding cold injury will, to a degree, apply to all of the stone fruits. As a group, they are more vulnerable to winter injury than pome fruits and consequently are grown more selectively.

Let's look now at the stone fruits in the order in which they ripen as we go through their care and growing requirements.

6. Sweet Cherries

Of the stone fruits, and actually of all the tree fruits, the sweet cherry is in most cases the first to be harvested as summer gets into full gear. Sweet cherries can be classified into two different groups fairly easily based on their shape and the firmness of their flesh.

The first group is the "Heart" group. The sweet cherries in this group are generally heart shaped (as opposed to round) and have fairly soft flesh. You won't normally find cherries from this group in the grocery store. Due to their soft flesh and short shelf life, they are not grown much by the commercial cherry industry. Several varieties of this group are well suited to the home garden, though.

More common, due to their popularity with commercial fruit growers, are the sweet cherries of the "Bigarreau" type. The fruit found in this group tend to be round in shape and have crisper, firm textured flesh.

Some varieties in each group have reddish juice and are considered "dark sweets". Those with pale almost colorless juice are often known as "light sweets".

Growth Habit and Training

The sweet cherry's natural tendency is to grow as a central leader tree. It will grow overly tall, if allowed to, and so is often trained to a modified leader system as discussed in Chapter 11. Sweet cherry trees can be fast, vigorous growing trees. For this reason it is desirable to leave more distance between scaffold limbs than would be done with other fruit. Generally twelve to fifteen inches of vertical distance works well. Lateral or side branches of the main scaffolds on sweet cherry trees tend to grow in close whorls of four or five branches. These whorls should be thinned

Figure 24. Sweet cherry types

trimming

back to two or three branches soon after they form. If left to grow for several seasons, the later pruning cut can have a stunting effect on the remaining shoot growth. Cutting back the tips of vigorous growing scaffold limbs each spring will help encourage more uniform lateral branch growth and fruit spur formation, too, while containing the tree to its allotted space.

45° angles

When the tree is young, it is very important that you train the branches to good, wide open forty five degree angles. Spring type clothespins are the preferred spreader for this purpose, as discussed in Chapter 13. Without spreading, narrow crotch angles on sweet cherries can be extremely susceptible to cold injury as well as being more brittle and quick to split if the fruit load is heavy.

Very often one branch of a young sweet cherry tree will be much stronger growing than the others. As it will rapidly dominate the other branches, you should subordinate it by cutting it back as needed, or if possible, by removing it completely. Otherwise a lopsided, nonsymmetrical tree can easily result and be harder to manage in later years.

Flowering and Pollination

Many parts of the country have celebrated the cherry for its blossom and, like the ornamental flowering cherry, the sweet cherry tree can be an aesthetic as well as a flavorful addition to the landscape. In order to garner a harvestable crop, though, you will need to plant at least two different varieties, since sweet cherry is self-unfruitful and does require pollen from an outside source in order to produce a crop. To challenge you a bit more, there is also some cross-incompatibility between certain sweet cherries. That is, not all varieties of sweet cherries will pollinate all other varieties. To add even more confusion, two varieties of sweet cherry, Lapins and Stella, are self-fertile and do not require cross-pollination by another variety. The simplest way to avoid a cross-incompatible combination is to consult the chart in Figure 25 or the nursery catalog you are ordering from.

Tart cherries are biologically capable of pollinating sweet cherries, but normally their time of bloom does not overlap well enough for tart cherry to be counted on as a pollen source for sweet cherry trees.

Most of the fruiting buds of sweet cherry are borne on short spurs off the main branches. The fruit bud itself is reasonably hardy;

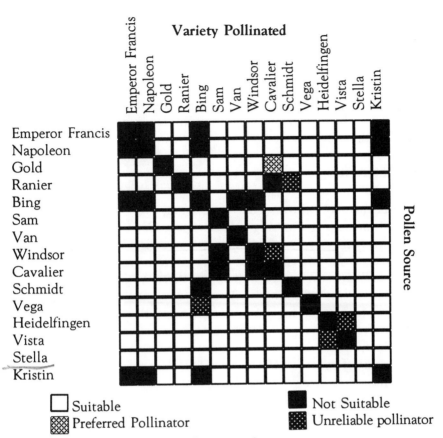

Figure 25. Sweet cherry pollination choices

comparable to domestic plum. Due to its early blooming nature, though, a sweet cherry tree has a period of time in the spring when the blossom bud is very easily damaged by frost. This period occurs in the few weeks just before bloom, typically in late March to early April on the west coast and mid to late April in the Great Lakes growing areas. At this time the flower bud cells are swollen with water and are commonly known as being in the "water stage". Should temperatures drop below about 28° F, the water freezes and expands inside the cells, often breaking the cell open and causing irreparable damage to the pistil of the flower. If cut open and examined, this frost damaged flower part will appear a translucent brown shortly after damage and will eventually wither and turn black. Since the cherry bud is so sensitive to cold at this critical period, cherries are grown most successfully in areas where the spring temperature fluctuations are tempered by large bodies of water. Some of these areas include Washington state, Wisconsin's Door peninsula, the Lake Michigan and Lake Ontario shorelines and Utah's Great Salt Lake area. One might think that by growing cherries in the southern United States you would escape the frost problems, and you would, but the high summer temperatures and humidity make cherry diseases difficult to effectively control there.

Rootstocks

One element of growing cherry trees that is simpler than growing other fruit trees is choosing the rootstock. Currently there are only two cherry rootstocks in widespread use in the United States. They are Mazzard and Mahaleb. The Mazzard rootstock is suitable for both sweet and tart cherries. Not all sweet cherry varieties are completely compatible with the Mahaleb rootstock. Most grow better on Mazzard root; so it has become the rootstock of choice for sweet cherry trees.

Mazzard - This rootstock produces a large tree approximately twenty five feet tall at maturity. Vigorous and productive, Mazzard rootstock is resistant to root knot nematodes and fairly tolerant of soil moisture. One selection of Mazzard rootstock named F12-1 has shown some resistance to bacterial canker and is often recommended in areas where this disease is a problem.

Mahaleb - Mahaleb rootstock does best in sandy loam soils and is not tolerant of truly wet soils such as a heavy clay. It produces a slightly smaller tree than Mazzard root, but not one small enough to be considered a dwarf by any means.

Colt - Although currently there are no true dwarf rootstocks for cherry trees, Colt is a step in that direction. It is relatively new and supplies are limited, so it may be somewhat difficult to find. Propagated in England, Colt produces a tree that is about 80% the size of Mahaleb. It shows some resistance to bacterial canker and crown gall. Colt is also more tolerant of wet soil conditions than the other cherry rootstocks. If you have limited space, growing a less vigorous scion variety propagated on Colt rootstock could be an option for you.

GM-61 - A newly released rootstock from Belgium, GM-61 is even more dwarfing than Colt. It produces a tree about fifteen feet tall and is hardy to -20° F. Since it is relatively new, it is still somewhat unproven under a wide range of growing conditions, but appears to offer hardiness and dwarfing that has been unavailable in cherry rootstocks until now. It is worthy of trial on a small scale.

Variety Selection

For home use, sweet cherries are usually eaten fresh, frozen or canned. All of the sweet cherry varieties are generally suitable to all of these uses. You will just have to decide if you prefer a dark or clear juiced cherry. Or, since two varieties are needed for pollination, try one of each. It is often hard to tell from a catalog which variety is which, the most adaptable and available ones are listed below.

Light Sweet Cherries

Emperor Francis	Napoleon (Queen Ann)
Gold	Ranier

Dark Sweet Cherries

Bing	Schmidt	Vega
Cavalier	Stella	Vista
Hedelfingen	Van	Windsor
Sam		

Sweet cherry growers are challenged by something that few growers of other fruit ever experience: as the sweet cherry nears harvest, a large portion of its juicy flesh is made up of water surrounded by a fairly inelastic skin. If rain occurs during this stage, the skins of many sweet cherry varieties tend to crack and split. Cracking seems to be worst when warm temperatures accompany or follow the rain. Not much can be done to prevent the cracking, unfortunately. If possible, pick the fruit soon after the rain and refrigerate it to keep fungus diseases from developing at the crack sites. If the fruit is not ready for harvest yet, it is a good idea to protect it with a fungicide spray, because warm wet conditions encourage rapid disease development. Certain varieties are more susceptible to cracking, and some varieties even react differently, depending on where they are grown. The chart below shows the varieties generally most resistant to cracking. Local experts may be able to advise you of others that do well in you area.

Varieties that Resist Cracking

Corum	Hedelfingen	Van
Emperor Francis	Lapins	
Gold	Sam	

Like most stone fruit, sweet cherries are not particularly tolerant of cold climates. A few varieties, listed below, are more likely to produce a crop in these areas. This is largely due to the fact that they are normally quite prolific producers. In milder climates though, they often set such heavy crops that small size fruit results. So choose them for situations where you otherwise would not be able to grow a reliable cherry crop.

Hardy Sweet Cherries

Blk. Republican	Gold	Windsor
Emperor Francis	Rainier	Yellow Glass
	Van	

Sweet Cherries

Average years to bearing:	5 to 6 years
Average yield per tree:	50 to 100 pounds on a mature tree
Space needed per tree:	20 foot circle
Average mature height:	25 feet
Days from bloom to harvest:	50 to 60 days
Pollination requirements:	Need cross pollination, except Lapins and Stella
Most common pruning system:	Modified central leader
Commonly used rootstock:	Mazzard
Common insect pests:	Curculio, cherry fruit fly
Common diseases:	Brown rot, leaf spot, bacterial canker
Disease resistant varieties:	None
Popular varieties & best use:	Emperor Francis (light) Van (dark), both used for eating and canning
Useful Life:	20 years

Number of trees for a family of four: 2 trees

Bear fruit on spurs that are 2 to 5 years old and on basal portion of new wood

I LOVE YOU

I ♡ U
I Love you

7. Tart Cherries

In the United States, tart cherries (sometimes also called sour cherries) are grown largely for use in cherry pies or for processing into other bakery and dessert items. The tart cherry tree is slightly smaller than a sweet cherry tree, with more fine lateral branching. Like the sweet cherry, it bears fruit on numerous fruiting spurs. Additionally, it bears fruit buds on one year old lateral (or side) branches. It is important to keep a tart cherry tree growing vigorously - studies have shown that if less than seven inches of new branch growth is made in a season, that wood will bear almost all fruit buds. In the following year the wood will become nonproductive, since no vegetative buds were borne to continue renewal of one-year wood. On bearing age trees, moderate annual pruning will also encourage continued production of fruiting wood.

Growth Habit and Pruning

Experience has shown that, although the tart cherry has a natural growth habit that is open and spreading, it benefits most from being trained to a modified central leader. In training the young tart cherry tree, though, you will want to treat it somewhat differently than other fruit trees that are trained to this system. After planting, remove side branches as you do with all whips, but *do not* cut back the leader. When new side branches grow, starting at about thirty inches from the ground, select and keep a scaffold limb on the southwest side of the tree or one facing into the prevailing wind. At about six inch intervals up the trunk, continue to select additional scaffold limbs in each of the compass directions. At least three, and preferably four or five scaffolds should be evenly distributed around the trunk. Then cut the leader to an outward growing lateral to form the open centered top of the tree. Prune only as little as necessary to select the desired branching structure in the early years or you will cause the tree to postpone bearing for several years.

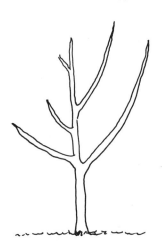

There are several things to avoid when training the tart cherry tree. First, try not to make any heading back cuts on branches that are intended to be part of the main scaffold system. Heading back cuts have been shown to cause a definite stunting of growth in tart cherry trees. Second, do not select two branches growing out of the trunk at the same height. Having two opposing branches of this nature often chokes the trunk and further branch growth above it. Also avoid allowing two branches to grow parallel and directly over each other. This can also cause stunted growth in the tree.

Figure 26. Stunted growth caused by parallel growing branches

Rootstocks

As already mentioned, rootstock choices for cherry trees are limited in number, making selection easy. The Mazzard stock discussed for use with sweet cherries is also suitable for use with tart cherries. Mahaleb cherry rootstock is probably the most widely used choice. It produces a slightly more compact tree than Mazzard, is more productive, and has better winter hardiness. Since it is more tolerant of drought conditions, it is well suited to the sandy soils common in the major cherry growing regions of the United States. Mazzard rootstock does not tolerate wet soils well, however. As with sweet cherry, Colt, a semi-dwarfing rootstock, is a possible choice if your space is limited.

Variety Selection

Although there are a number of tart cherry varieties, only a limited few make up the selection available to the gardener or professional grower. All tart cherry varieties are self-fruitful, so one average size tree alone will still give you a bountiful crop.

Tart cherries can be broken down into two groups. The group grown most commonly in the United States has bright red skin, a translucent yellow flesh and clear, almost colorless juice. They are sometimes referred to as being of the amarelle type.

Cherries of the Morello type are more commonly grown in Europe. They have a bright red flesh and dark red juice, making them very attractive for pies. Some people claim they are slightly sweeter, too. The most available tart cherry varieties, in order of ripening, are described below:

Early Richmond is an early ripening variety that is harvested about ten days ahead of Montmorency. Its fruit is smaller in size and the flavor more acidic. It typically yields less than the same size Montmorency tree.

Morello is a dark juiced cherry which makes a very attractive pie. Maturing just before Montmorency it is able to hang on the tree longer without loosing quality. It is quite susceptible to cherry leaf spot fungus.

Montmorency is by far the most widely available tart pie cherry variety found in nursery catalogs. Bright red in color, it has a yellow flesh and pale, almost colorless juice - much like a "light" sweet cherry. It ripens in mid to late July in the major Great Lakes cherry producing regions.

Meteor has a light red flesh with red juice, although not as dark as the Morello. It ripens just after Montmorency. The Meteor tree has a more spur type growth habit than Montmorency and in some areas will grow only 80% to 90% the size of Montmorency. In addition, it is quite cold hardy.

North Star is another hardy tart cherry variety that has the added benefit for the backyard fruit grower of being a natural dwarf tree as well. Consider this tree for a far northern location, as the original parent was found in Minnesota. Like Meteor, its fruit is red fleshed with red juice with harvest occurring in late July to early August.

Average years to bearing:	4 to 5 years
Average yield per tree:	30 pounds at around 7 years 50 to 100 pounds at maturity
Space needed per tree:	15 to 20 foot circle
Average mature height:	20+ feet
Days from bloom to harvest:	60 days
Pollination requirements:	Self-fertile, only Morello benefits from cross pollination
Most common pruning system:	Modified central leader
Commonly used rootstock:	Mahaleb
Common insect pests:	Curculio, cherry fruit fly
Common diseases:	Brown rot, leaf spot
Disease resistant varieties:	None
Popular varieties & best use:	Montmorency - pies, cooking
Useful Life:	15 years

Number of trees for a family of four: 1 tree

Bears fruit on spurs and one year old wood

8. Apricots, Plums, Apriums & Pluots

Apricots and plums are so closely related botanically, that many of the rootstocks used are common to both species. Much of the care you will give them will be similar too, and now that some interesting crosses of the two - apriums and pluots - are becoming available, they will have even more in common. Consequently, to avoid repetition, we have grouped them together in this single chapter.

Apricots

The apricot is a large and sturdy growing tree. Apricot fruit buds are similar in hardiness to peach but quite susceptible to drying by cold winter winds. Consequently, apricots produce best if grown in a sheltered spot in the landscape. The apricot is the earliest fruit tree to flower in spring and thus it does not always produce a large crop since the blossoms are often affected by spring frosts. The blossom is a beautiful white, though, and the tree is a sight to enjoy when in full bloom. It can be a stunning accent in the landscape even in years when fruit production is reduced by frost.

Growth habit of the apricot tree is similar to sweet cherry, and it is best trained to a modified central leader system as discussed in Chapter 11. As with most fruit trees, apricots should be pruned lightly in their early years, just enough to develop a sufficient scaffold structure. Mature trees benefit most from a moderate annual pruning in early spring. In northern regions, you may want to postpone pruning until after bloom and spring frosts. This way you can tailor your pruning severity somewhat to the amount of crop you anticipate. You will want to direct much of the pruning of your mature apricot tree to keeping it contained to the height and width practical for your garden.

Flowering and Pollination

Apricot trees tend to bloom and set fruit abundantly. Most apricots are self-fertile and require no additional pollinators. The fruit buds are normally produced on spurs that fruit for two to four years and on the tips of last season's shoot growth. Regular moderate pruning serves to renew adequate fruit producing wood. Hand thinning of apricot fruit does not seem to benefit the fruit in the same way as it does some other fruit. Pruning has been shown to be the best way to reduce competition among excess fruit.

Rootstock

Apricot seedlings and plum are both used as rootstock for apricots. When soil conditions allow, it is advised to bud apricots to apricot rootstock. Several possibilities are available for budding plum on a number of the stone fruit roots, including apricot, almond, and peach. The most successfully used choices include:

Manchurian hardy apricot - This seedling apricot rootstock performs best in light, sandy soils. It is the most reliable and compatible rootstock for use with apricot. It produces a large, vigorous tree.

Myrobalan - This seedling plum rootstock is very well adapted to heavy, wet soils. It produces a well anchored, standard size tree. Some strains of this rootstock are also nematode resistant. Myrobalan plum is the most frequently used to grow plum varieties and can be used as a rootstock for apricots if your soil is not suitable for apricot seedling rootstock.

Marianna 2624 - Another plum rootstock, Marianna 2624 is not as well anchored as the preceding choices. It is however resistant to nematodes, oak root fungus, and tomato ring spot virus. Marianna can be a particularly good choice in areas of the Great Lakes states and New England where tomato ring spot virus is a common problem.

St. Julian - A series of somewhat dwarfing plum rootstock, St. Julian is well anchored and suited to heavy soils. It is more difficult to propagate than some of the other plum rootstocks, so may be more difficult to find. The St. Julian rootstock is actually a series of selections. Minor differences exist between each selection in the series so ask you nursery about the selection they are using. St. Julian A is the most common.

Pixy - A newly named selection from the St. Julian series, Pixy produces a tree that is about 50% of standard size. It is very productive, but quite sensitive to drought, so adequate irrigation is needed. The rootstock is somewhat difficult to propagate and consequently hard to obtain, but worth trying if a dwarf tree is desired. It is not compatible with apricot.

Peach rootstocks - Peaches and plums are botanically closely enough related that some peach rootstocks are successfully used with plums as well. In areas where nematodes pose a serious problem, Marianna plum can be used as can some of the nematode resistant peach stocks such as Nemaguard. If using peach rootstocks, however, be aware that they will not tolerate wet conditions the way plum stock will. More information on peach rootstocks can be found in Chapter 9. Your nursery can also advise you regarding the best choice they have available.

Varieties

The number of available apricot varieties is small compared to most other tree fruit. Certain varieties seem to be specifically suited to growing in certain areas of the country. Some varieties do not like extremely warm temperatures, but at the same time are not cold hardy either. Others thrive in warm areas. Some of the most reliable choices and their respective regions are as follows:

Regionally Adapted Apricot Varieties

South	Moorpark
Pacific Coast	Blenheim, Goldbar, Goldrich, Goldstrike, Moorpark, Perfection, Puget Sound, Rival, Tomcot, Wenatchee
Great Lakes **Northeast**	Goldcot, Traverse, Curtis

Winter Hardy Apricots

Alfred	Hardy Iowa (late	Moongold*
Curtis	flowering)	Sungold*
Goldcot	Hargrand	Traverse
Harcot	Harlayne	

*need to be planted together to pollinate each other

Low Chill Apricots

Earligold
Garden Annie (600)
Newcastle (350)
Perfection (600)

High Chill Apricots

Scout (1000)
Tilton (1000)

Plums

Of all the stone fruit, plums offer potentially the most variety and improved quality when grown in the home orchard. Only a very small selection of the over 2000 plum varieties to choose from is used by the commercial plum industry. Most of those are used for drying as prunes, canning or are shipped immature and firm to withstand the long distance travel. Unfortunately, this often results in a fruit that doesn't offer nearly the juicy flavor it could. As a home orchardist, though, you can benefit greatly from the flavorful choices available when the plum is left to ripen on the tree!

Plums are generally broken into four major categories: the native American plums, the European plums, the Japanese plums and the Japanese-American plum crosses.

American Plums

The American plums are known by a number of other names as well: sand cherry, beach plum, chickasaw or Canada plum. This group of plums tends to be very tart, generally too tart for pleasurable eating, but excellent for making jams and preserves. As a native plant, this group as a whole is quite hardy and fairly disease resistant. For these characteristics, they have often been used in plum breeding programs. Some of the varieties can be quite thorny or grow as more of a shrub form than a tree. The fruit is smaller in size than most of the other plum types, usually around one inch in diameter.

European Plums

Most of the commercially important varieties and many of those that you will find in the nursery catalogs fall into the European plum category. In the United States, most of the varieties from this group are used in commercial processing, but in Europe they are commonly eaten fresh. Because of the wide variation of characteristics in this group, it is further broken down into:

Prune plums - used mostly for commercial drying as prunes. These oblong shaped fruit can usually be dried without the need to remove the pit.

Blufre	Italian	Verity
Earliblue	Stanley	Vision
Imperial Epineuse	Veeblue	

Gage plums - Sometimes also called Reine Claude plums, this group is composed of a number of varieties suitable for fresh eating or canning. The sweet fruit is generally round in shape with tender, juicy flesh. The color is normally yellow, green or light red. They generally grow best in a warm, sunny climate.

Bradshaw	General Hand	Pearl
Bryanston Gage	Green Gage	Reine Claude
Count Althan's	Imperial	Seneca
Empress	Iroquois	
French	Morentini	

Blue plums - Sometimes known as Imperatrice plums, the fruit in this large group are almost all blue, of medium size and oval in shape. Most have a heavy powdery white "bloom" over their thick skin. The flesh is usually firm and of fair eating flavor.

Early Laxton	Mount Royal	President	Valor

Lombard plums - The fruit in this group are usually somewhat smaller in size than the blue plums and red in skin color. Many of the varieties are of fair eating quality as well.

Yellow Egg plums - Normally small and yellow in color, this group has only a few varieties, most of which are primarily suited to canning.

European plum trees are usually vigorous, upright growing trees with smooth, grey bark. Leaves are thick, dark green and smooth on top with a pale green, hairy (pubescent) surface on the underside. Leaf edges have coarse notches. Most of the fruit is borne on spurs. The fruit may have either a freestone or cling stone.

Japanese Plums

Japanese plums are normally large in size, heart or oblate shaped, and bright red, yellow or purple in color. Many have a nice desert eating quality with yellow, golden or red firm, juicy flesh.

The Japanese plum tree's growth habit can vary from upright to spreading. The leaves are normally smooth, sharply pointed on the ends, and smaller in size than those of domestic plum. The bark is rough textured in contrast to the European varieties. Fruit and the profuse bloom of the Japanese plums are borne on both spurs and one year old wood. The trees tend to bloom earlier in the season and be more frost susceptible than the European or native cultivars. Least hardy of the plums, Japanese plums' winter hardiness tends to be similar to peaches, so they are normally considered suitable for areas where peaches can survive. Some of the common varieties include:

Abundance	Early Golden	Red Heart
Ace	Elephant Heart	Royal Garnet
Black Amber	Formosa	Santa Rosa
Burbank	Mariposa	Shiro
Burgundy	Methley	Simca Plum
Carolina Harris	Ozark Premier	Vanier
Cocheo	Peach Plum	Wickson

Japanese-American Hybrids

The Japanese-American plum crosses have been bred to take advantage of the disease resistance and hardiness of the native varieties while incorporating the eating quality of the Japanese cultivars. These varieties may serve you well where the others are lacking characteristics needed for your growing area.

Ember	Redcoat	Toka
LaCresent	South Dakota	Underwood
Pipestone	Superior	Waneta

Pollination and Thinning

Although a few of the European varieties are self-fruitful, the large majority of plum varieties require cross pollination, so two or three varieties should be planted together. The Japanese varieties tend to bloom earlier in the season than the European types and therefore may not be reliable pollen sources for them. The Japanese cultivars will serve as adequate pollen source for the Japanese-American hybrids. The hybrids will also pollinate the Japanese varieties, although not well, nor will the European strains serve as good pollinators for the Japanese plums. In choosing plum varieties, pay attention to both the type of plum and the timing of bloom.

Most plums have a strong tendency to set much more fruit than the tree can reasonably support and size adequately. This can cause plums to develop a biennial (every other year) bearing pattern if not thinned after June drop. When the fruit has grown to about the size of your thumbnail, hand thin to one fruit per cluster or spur with a minimum of five inches between fruit. This will help the tree produce annual crops of large size fruit.

Rootstocks

The most commonly used rootstocks for plums have been discussed earlier in this chapter. A mention about dwarfing rootstocks for plums should be made here though, since home orchardists may feel they need a dwarf tree for their limited space. Although Western Sand Cherry (*Prunus besseyi*) and sand plum (*P. angustifolia*) have been advertised as dwarfing stocks for plum, both have been shown to have incompatibility problems with many of the plum scion varieties and often suffer graft union breakage or tree death after several years of growth.

Plum trees are naturally medium size when mature; therefore a properly pruned plum tree can be kept reasonably contained. It is wiser to stick with a standard size plum rootstock and insure a long lived, productive tree or try the newly released Pixy rootstock.

Pruning

The growth habit of the plum falls into one of two categories: naturally upright varieties suited to a modified central leader training system; and open, spreading varieties, such as the Japanese types, that are trained to an open center. When training plums to an open center, it is generally recommended that four or five main scaffold branches be developed as opposed to the customary three used with peaches. Several additional secondary scaffolds are often maintained as well. More detail on both these training systems can be found in Chapter 11.

Young plum trees should be encouraged to grow twelve to twenty four inches of new shoot growth in the early years. In the bearing years, shoot growth of twelve to fifteen inches is adequate to encourage development of fruiting wood. Be aware that heavy pruning combined with heavy fertilizing can encourage considerable vegetative growth of water sprouts in plums. This can sometimes be controlled by bending the branches back down and tieing them to the trunk in an old practice known as festooning. As with most fruit trees, a moderate annual pruning is best for European plum varieties. The Japanese plums, which grow more vigorously, should also be pruned annually, but can be pruned more aggressively.

Plum fruit buds develop in two locations: one year old lateral wood and older fruiting spurs. Pruning should be done with the aim of getting adequate light to these fruiting areas, especially on the inside of the tree. After several years of heavy fruiting the spurs may become weak and less productive. Fruiting laterals may also become weak and brittle. It is a good idea to head back to vigorous laterals and remove depleted fruit spurs on a regular basis.

Apriums & Pluots

Several interesting crosses between apricots and plums have been developed recently. They are known as apriums, pluots, or plumcots. Most of the work of developing these fruit for the home garden has been done by long time plant breeder, Floyd Zaiger.

Basically, apriums more closely resemble apricots and require an apricot as a pollinator. Flavor Delight is the most common variety.

Pluots are 75% plum and 25% apricot in their genetic makeup. Plumcots have 50-50 plum and apricot parentage. Either an aprium or Japanese plum can serve as a pollinator for these trees. Several varieties are available:

Flavor Queen - a yellow skinned and yellow fleshed fruit, that has a sweet apricot after taste.

Flavor Supreme - a red fleshed fruit, that ripens early in the season. Has a flavor like Elephant Heart plum.

Mesch-Mesch-Amrah - this fruit has a unique raspberry flavor, which is more pronounced in some seasons than others.

QUICK REFERENCE
Apricots

Average years to bearing:	4 to 5 years
Average yield per tree:	50 to 100 pounds
Space needed per tree:	20 foot circle
Average mature height:	20+ feet
Days from bloom to harvest:	80 - 90 days
Pollination requirements:	Many are self-fruitful, but produce better with cross pollination
Most common pruning system:	Modified central leader
Commonly used rootstock:	Apricot seedling
Common insect pests:	Curculio, oriental fruit moth
Common diseases:	Brown rot
Disease resistant varieties:	Harcot, Harogem
Popular varieties & best use:	Depends on region of the country, see chart page 97
Useful Life:	12+ years

Number of trees for family of four: 1 to 2 trees

Bears fruit on 2 to 4 year old spurs and tips of new growth

Plums

Average years to bearing:	3 to 5 years
Average yield per tree:	40 to 60 pounds
Space needed per tree:	15 foot circle
Average mature height:	15 to 20 feet
Days from bloom to harvest:	75 - 120 days, depending on variety
Pollination requirements:	Most varieties require cross pollination, be sure to use European varieties to pollinate European varieties and Japanese to pollinate Japanese
Most common pruning system:	Modified central leader or open center
Commonly used rootstock:	Myrobalan, St. Julian X
Common insect pests:	Curculio, apple maggot, mites, oriental fruit moth
Common diseases:	Brown rot, black knot, bacterial spot
Popular varieties & best use:	Varies regionally with adaptability
Useful Life:	15 years

Number of trees for family of four: 2 trees

Bears fruit on one year old lateral wood and older spurs

9. Peaches & Nectarines

One of the oldest known fruits, peaches were grown in ancient China as long ago as 2000 B.C. Today they are grown in many parts of the world, most commonly in the areas of 25 to 45 degrees latitude both north and south of the equator. We will discuss the peach and the nectarine collectively as one group because genetically they are similar. The only major difference is the "fuzz" on the skin of the peach which is absent on the nectarine.

Growth Habit and Pruning

Peaches and nectarines are commonly the shortest lived of the tree fruits. Often succumbing to borers or winter cold damage, the peach tree's normal life is usually only ten to fifteen years. This makes it particularly important to get the tree's framework off to a proper start. Your reward will be a tree that starts to bear a crop in its third or fourth year.

Peach and nectarine trees are normally pruned to an open vase, as discussed in Chapter 11. Right after planting, select three strong wide-angled branches that are well spaced around the trunk. Remove all other branches and cut the three selected scaffolds back to two to three buds each. Whenever possible, have one of the three scaffolds growing into the prevailing wind. Try also to have about six inches vertical distance between scaffolds on the trunk. Develop the open center from these three.

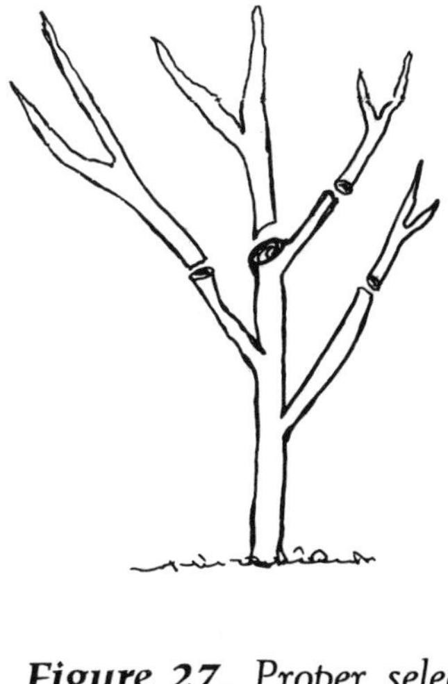

During the next two to three years, prune the peach tree as lightly as possible. Retain the scaffolds selected in the first season. Thin

Figure 27. Proper selection of peach branches

out only branches that grow from the trunk in competition with your selected scaffolds, those that grow down, straight up or across the center of the vase. Once the main scaffolds are about thirty to thirty six inches long, light heading back cuts can be made to encourage side branching.

From the third through the sixth year of growth, light annual pruning to continue shaping the open vase is the best course to pursue. By the sixth year, the tree should be around twelve feet high and wide and producing a regular crop. From this point on, it is important to prune the peach or nectarine tree every year, removing 20% to 40% of the old bearing surface each year. New growth of eight to twenty four inches long will be the most productive.

The best fruit is normally produced in the upper third of the tree, and annual pruning will keep the fruit bearing branches within manageable reach. This pruning also reduces the amount of hand thinning necessary, as the peach usually produces many more fruit buds than are needed for a full crop.

What time of the season to prune is always a question that stumps the new fruit grower. With peaches, the answer will depend more on what your location is than with other fruits. Most peach pruning is done in late winter or early spring. If you live in a northern climate where the tree may be subjected to winter temperatures of -20° F or below, try to delay pruning until late spring. At this time, freeze damage to both fruit buds and branches can be assessed more easily. Then, prune accordingly. Remove wood that was freeze damaged and prune moderately in the portion of the tree that still has live wood and buds.

If you live in an area with a more moderate winter climate, frost damage to the opening blossoms may be your major problem. Make some thinning out cuts in late winter and possibly a few more after bloom, when the danger for frost damage to the flowers has passed.

Flowering and Pollination

Peaches and nectarines normally produce an abundance of fruit buds. The buds typically occur on lateral wood grown the previous season. When looking at a branch, the fruit buds will appear as the two plump buds that surround a smaller vegetative bud. Most peach and nectarine varieties are self fertile, requiring no other pollinator. (A few exceptions exist, but are not widely used.) You may, however, enjoy planting several varieties of peach or nectarine to lengthen the period over which you have fresh fruit to eat.

Rootstocks

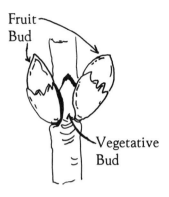

Figure 28. Peach buds

Rootstock for peaches and nectarines offer the backyard fruit grower several options, depending on your location. All peach rootstocks grow best in light textured, well drained soils. Problems that have occurred with peach rootstocks in the past include susceptibility to nematodes, viruses and bacterial diseases. Cold injury also poses a problem in some areas. With varied characteristics, the rootstocks below are all being used with reliable results and should provide you with a choice to match your climate.

Bailey - This relative newcomer to the collection of peach rootstocks originated in Iowa. It shows promise as a cold hardy stock. Resistant to root lesion nematodes, Bailey rootstock produces an abundant root system and a standard size tree.

Halford - A seedling offspring of the Lovell rootstock, Halford is adaptable to a wide range of conditions and is very similar to Lovell in its characteristics. It produces a standard size tree.

Lovell - The most widely used peach rootstock today, Lovell has higher disease resistance than most other peach rootstocks. It produces a standard size tree that is compatible with most peach and nectarine varieties. It is not nematode resistant.

Nemaguard - Nemaguard rootstock is very resistant to root lesion and root gall nematodes, which can be a problem when peaches are replanted on an old peach site. Its high tolerance of heat and drought make it popular in the southern United States.

Citation - A peach/plum hybrid developed in California, Citation is the first true dwarfing peach rootstock. Peach trees grown on it are about 50% of standard size. Its use has been limited because it is not particularly winter hardy and also has some incompatibility problems.

Varieties

Peaches and nectarines can be broken into two basic groups: clingstone and freestone varieties. As the name suggests, cling stone varieties have fruit in which the fruit flesh adheres or "clings" to the pit when cut open. The flesh of clingstone varieties tends to be firm and smooth textured when canned. Syrup from these varieties may seem thicker too, due to their higher soluble pectin content.

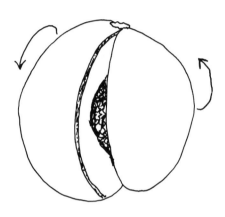

Figure 29. *Releasing the pit in a freestone peach*

Freestone varieties will normally allow the pit to be removed without flesh clinging to it when the fruit is ripe. The easiest way to release the pit is to cut the fruit in half from the stem end to the blossom end. Gently but firmly turn the two halves in opposite directions. The pit will normally pop away from one half and can then be gently pried from the other half. The fruit flesh will normally be softer and have a slightly stringy texture when canned. From the standpoint of growing and caring for the peach tree, there is no difference between the two. If, however, you want to use most of your harvest for canning or cooking, you may want to select freestone varieties for convenience.

In general, varieties ripening early in the season tend to be cling stone while later ripening varieties are freestone. The ripening date of the Redhaven peach is commonly used as a reference date to indicate harvest time. Those varieties maturing before and up to a week later than Redhaven are termed early season varieties. Mid season varieties mature up to two weeks later than Redhaven; late season varieties follow, ripening up to a month after Redhaven. Many catalogs will indicate relative ripening date with a number and plus or minus sign in parentheses behind the variety name. The notation Candor (-19) means the Candor peach typically ripens nineteen days before Redhaven would in your area. Most yellow fleshed peach varieties are reasonably suited for both fresh eating and canning. Select varieties most suited to your climate and according to ripening season as well as your personal taste preference. Nectarines are grown primarily for fresh eating. Varieties can be selected in the same way that peaches are. Be aware, though, that nectarines tend to be more susceptible to winter cold injury.

White Fleshed Peaches

Peaches are sometimes also broken down into groups by flesh color. Most peaches are yellow fleshed, but a few are white fleshed. White fleshed peaches have been a novelty on the market, but are very worthy of consideration by the home fruit grower. Often neglected by the commercial grower because the cream colored flesh shows bruises very easily, white fleshed peaches are intensely flavored, with a sweetness that is unsurpassed. They are best suited for fresh eating, straight from the tree.

The white fleshed peach varieties are often more bud hardy and better able to survive low temperatures than some of the yellow fleshed varieties.

White Fleshed Peaches and Nectarines

Babcock	J. M. Mack	Peregrine
Belle of Georgia	Karla Rose	Raritan Rose
Champion	Melba	Red Chief(N)
George IV	Oldmixon Free	Springtime
Gold Mine(N)	Pallas	

* (N) denotes nectarine varieties

Peaches for Special Situations

Winter Hardy Tree

Belle of Georgia
Jefferson
Madison
Oldmixon Free
Pocahontas
Raritan Rose
Reliance

Winter Hardy Fruit Buds

Belle of Georgia
Earliglo
McKay
Oldmixon Free
Raritan Rose
Reliance

Low Chill Varieties

Bonita (350)	Melba (500)
Desert Gold (200-300)	Mid Pride
Fantasia(N) (500-600)	Silverlode(N)
Flamecrest(N) (500-600)	Sungold(N) (500)
Flordasun (300)	Sunred(N) (300)
Gold Mine(N) (500-600)	Tejon (400)
Gold Dust (500)	Ventura (400)

* ( ) numbers denote number of hours of chilling required

Leaf Curl Resistant Varieties

Dixired   Indian Free   Loring

Need for Thinning

As mentioned earlier, the peach tree normally produces many more fruit buds than it is capable of maturing as full sized fruit. On average, only 10% of the fruit buds present need to grow to maturity to provide an ample harvest. This may partially be nature's protection again frost damage, but even when frost damage has occurred it may still necessary to remove some of the excess fruit from the tree after pollination.

Approximately two weeks after shuck split, you will be able to see what fruit is likely to continue growing and what will be shed in the "June drop": that fruit which is likely to mature will appear larger and will continue growing. Other fruit will remain smaller and seem to stop growing. If you carefully open these smaller fruit, you will see that the pit is really not forming inside. These small fruit are destined to drop from the tree naturally in the next week or two.

Your major concern now will be the fruit that continues to grow. In order for the fruit to reach a mature diameter of 2½ inches or more, the developing fruit should be thinned. A good time to thin peaches is around six to eight weeks after bloom in cooler climates and four to six weeks after bloom in warm locations. Simply pick off the excess fruit by hand.

There are two "rules of thumb" for deciding how many fruit to leave on the tree. With the first method you leave an average of six to eight inches of branch distance between each peach. Because it is easy to visualize, this method is most commonly used and easiest for the inexperienced. With the second method, keep in mind that each peach needs about thirty five leaves to support its growth to full size. Here you thin according to the vigor of the vegetative growth. This method can be most helpful if, for some reason (severe insect infestation or disease), the tree is in a weak growing state but has produced an abundance of fruit. You can use this guideline as a way to help the tree adjust to the stress of fruiting.

Some varieties tend to have naturally smaller fruit and should therefore be thinned more aggressively. Most nursery catalogs will indicate these as needing heavy thinning. Experience will be your best teacher. If you are growing your fruit without the use of chemical pest controls, this is also an opportunity to remove any that have been damaged by insects.

Split Pits

Occasionally a peach grower will find fruit that appears to be cracking open along the suture line or fruit in which the pit is broken into pieces inside the fruit. This condition, known as "split pits", occurs when insufficient moisture is supplied to the developing fruit early in the season followed by sudden abundant moisture when the fruit is rapidly increasing in size just prior to harvest. The best remedy is a steady natural water supply or supplemental irrigation.

A Few Tips About Nectarines

As mentioned, caring for nectarine trees is almost identical to caring for peaches. A few small, but important, differences do exist though; requiring a little bit more attention on the part of the nectarine grower. Each of these points will apply for nectarines as a group, although there may be exceptions for individual varieties.

1) Since they are "fuzzless", nectarines are slightly more susceptible to insect damage and rot diseases. Their skin also tends to crack more easily under adverse conditions.

2) Bacterial spot disease can be a more severe problem with nectarines.

3) Nectarines require slightly more nitrogen than peaches to produce a large, high quality crop.

QUICK REFERENCE
Peaches/Nectarines

Average years to bearing: 4 years

Average yield per tree: 2 to 2 1/2 bushels

Space needed per tree: 15 foot circle

Average mature height: 15 feet

Days from bloom to harvest: 60 to 120 depending on variety

Pollination requirements: Self-fruitful

Most common pruning system: Open center

Commonly used rootstock: Halford or Lovell

Common insect pests: Oriental fruit moth, tarnished plant
                                 bug, curculio

Common diseases: Leaf curl, valsa canker, brown rot,
                                 bacterial spot

Disease resistant varieties: See chart page 112

Popular varieties & best use: Redhaven - eating, canning
                                 Redskin - eating, canning

Useful Life: 12 years

Number of trees for a family of four: 2 to 3 trees

Bears fruit on one year old wood

Section III.

Caring for Your Fruit Tree

10. Nutrition and Fertilizers

All living plants need certain nutrients and energy to grow and fruit properly. Green plants are uniquely adapted to provide for their needs through the cycle of photosynthesis. During photosynthesis, the fruit tree takes in sunlight and uses its energy to produce the sugars it requires. A closer look will explain the process.

Photosynthesis

For proper photosynthesis, the fruit tree needs sunlight, water, carbon dioxide, oxygen and certain minerals. The roots absorb water, oxygen, and minerals from the soil below. These ingredients are transported through vessels in the trunk and branches up to the green leaves. Chlorophyll, a green pigment in the leaf, absorbs sunlight and transforms it into usable energy. This energy is then used to process carbon dioxide absorbed from the air and other nutrients transported from the roots. The result is sugar and other carbohydrates.

Some of the sugars produced are converted to energy and used right away by the tree to maintain its daily growth process. Unused sugar is stored for future use in various locations throughout the tree. Much of the sugar is stored in the fruit and accounts for its sweet taste. Figure 30 should help you visualize how this works.

You can see now why it is advantageous to care for the tree so that it can achieve maximum production and storage of sugar. Many factors affect the plant's ability to carry on efficient photosynthesis. You may understand better now why you should select a soil that is not so heavy and waterlogged that it holds insufficient oxygen. One of the main purposes of pruning is to allow sunlight into the tree. What you do in establishing an open tree structure will have a great

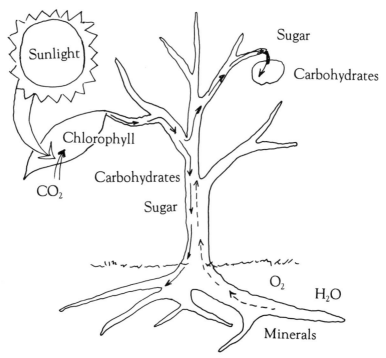

Figure 30. *The photosynthesis cycle*

effect on the amount of available light. Healthy, disease free leaves are best able to attend to photosynthesis. It is interesting to note that each individual growing fruit needs the support of as many as thirty leaves. Additional leaves are needed to support the growth processes of the limbs and roots. Future chapters will discuss pruning and control of diseases. In this chapter we will examine plant nutrients, their sources and why they are important to the fruit tree.

Choosing a Fertilizer

Much like vitamins, fertilizers come in many shapes and kinds. To make choosing a fertilizer easier let's look first at the label, which by law must be printed on every package of fertilizer. This label must give you the percentage of three major nutrients, nitrogen, phosphorus, and potassium. You will hear these referred to as N-P-K, in reference to their chemical element symbols. You may also see notations of numbers such as 15-5-10. These numbers indicate that the fertilizer is 15% nitrogen per pound of fertilizer, 5% phosphorus, and, 10% potassium. Fertilizers that include all three major nutrients are referred to as complete fertilizers. Percentages of other minor nutrients may also be indicated on the label. Later in this chapter you will learn how these numbers can help you calculate how much actual fertilizer to apply.

Function and Sources of Nutrients

Plant nutrients are commonly divided into 3 groups: major, secondary and trace nutrients. The major nutrients are nitrogen, phosphorus, and potassium. All three are crucial for plant growth.

Major Nutrients

Nitrogen - Nitrogen, as a major nutrient, is primarily responsible for shoot growth and green color in the leaves. The fruit tree uses nitrogen that is in the chemical form of nitrate nitrogen. Calcium nitrate (15-0-0), sodium nitrate, and potassium nitrate (13-0-44) are common fertilizers that supply this immediately available form of nitrogen.

A less readily available form of nitrogen is ammonium nitrogen. It can be found in ammonium sulfate (21-0-0), monoammonium phosphate (11-52-0), or diammonium phosphate (18-46-0). Over a period of several months, ammonium nitrogen is converted by chemical processes in the soil, to nitrate nitrogen. Another fertilizer, ammonium nitrate (33-0-0), provides both nitrate and ammonium nitrogen for combined quick and slow release of nitrogen.

A third form of nitrogen is organic nitrogen. In this case, the term organic is meant as its chemically primitive form - that carbon is included in its makeup - not the popular current connotation of "organic". Organic nitrogen must be converted first to ammonium nitrogen and then to nitrate nitrogen. This process takes time, sometimes up to several years to complete. Urea fertilizer (45-0-0) is a common form of organic nitrogen.

Nitrogen is normally in low supply and heavily used by the fruit tree. In most regions of the country it will be necessary to supplement the soil with a nitrogen source on a regular basis, usually every year. At the same time it is important to apply nitrogen judiciously. It can easily burn tree roots or cause excess shoot growth at the expense of producing fruit buds.

Phosphorus - The fruit tree relies on phosphorus to help in the growth of roots, seeds, and early leaves. Phosphorus is not used directly, but rather converted to nucleic acids for use in growth processes. Slow to work in the soil, since it is not readily dissolved, phosphorus should be applied to the soil ahead of planting when needed. It can be found in rock phosphate and complete fertilizers.

Potassium - Potassium, the third major nutrient, is used by the tree in producing fruit, growing roots and resisting disease. It works by helping to transport starch and sugar through the tree. It is found in potassium nitrate, potassium sulfate, and muriate of potash.

Secondary Nutrients

Calcium, magnesium, and sulfur are often called secondary nutrients. They are as important to tree growth as the major nutrients, but needed in much smaller quantities. Calcium nitrate fertilizer is a common source of calcium and epsom salt is often used as a source for magnesium. Both can be applied as a foliar spray or dry calcium nitrate is often broadcast on the ground as well. Secondary nutrients may need to be supplied to your tree on an occasional basis. Soil or tissue tests can be used to determine when and how much is needed.

Trace Nutrients

Trace nutrients are all necessary for plant growth functions but in very small, or "trace" amounts. An excess of trace nutrients can even cause a poisoning or toxicity to your fruit tree. The trace minerals are boron, copper, chlorine, iron, manganese, molybdenum, and zinc. Seaweed emulsion is sometimes used as a source of trace minerals. Deficiencies and excesses of these minerals vary considerably in different parts of the country. Tissue testing and consulting with your local extension agent will be the safest way to determine if and when additional trace nutrients are needed in your fruit tree planting.

Nutrient Sources

Fertilizer Material	Actual % Nutrient By Weight			
	N	P_2O_5	K_2O	Other
Rapid Release Rate				
Ammonium sulfate	20			24 S
Ammonium nitrate	32.5			
Calcium nitrate	15.5			20 Ca
Diammonium phosphate	17-20	46		
Muriate of potash			60-62	
Potasium nitrate	13.4		44	
Potassium sulfate			45-52	
Seaweed immulsion (foliar)	1	1	4	numerous trace
Superphosphate		20		20 Ca
Triple superphosphate		46		
Urea	42-46			
Wood Ash		1-2	3-7	
Moderate Release Rate				
Animal Tankage	7	10		
Bat Guano	5-12	8-11	2	
Blood Meal	12	2	.5	
Manure (varies with type)	.25-6	.15-4	.25-3	
Kaolinite			12	
Milorganite	5	2-5	2	
Monoammonium phosphate	11	52		
Potassium magnesium sulfate (SulPoMag)			22	11 Mg 11 S
Slow Release Rate				
Basic slag		3.5-8		3.5Mg
Bone meal		18.5		22 Ca
Colloidal phosphate		14-28		26-33 Ca 4 Fe
Compost	2	6	3	
Greensand			5-7	
Potassium carbonate			56	
Rock phosphate		20-32		32 Ca

Soil and Tissue Testing

Before applying nutrients to your planting site it is wise to know what nutrients are already present in your soil or tree. By testing the soil you can avoid spending money on unneeded or incorrect fertilizers. Also, by avoiding over-application, you avoid mineral leaching that contributes to ground water pollution. Two forms of plant nutrient testing can be helpful to the fruit tree grower. The first is soil testing.

Soil testing is quite simple. Samples are taken with a probe that can be pushed eighteen to twenty four inches into the soil. The probe is a metal cylinder about an inch and a half in diameter and about thirty inches long. One side is cut out along its length and a handle across the top allows you to push it into the soil and pull out a core. You may be able to borrow one from your county extension office to do your sampling.

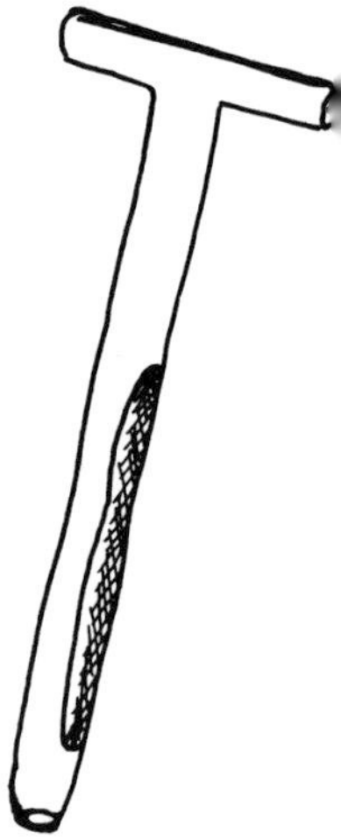

To take a soil sample gather fifteen to twenty sample cores, to a depth of around six inches, from your planting area. (For a homeowner with a small lot, this could include your whole backyard.) Mix them together well in a clean plastic bucket. Then bring a portion of this mixture (two cups is common) to your extension office in a plastic bag. The sample will be forwarded to a laboratory and you should have results back in three to six weeks.

Soil tests can tell you what nutrients are present in the soil, but if you are having a serious problem with growing your fruit tree, a tissue test can give you a more accurate measure of which nutrients have actually been taken up into the plant. The test can be fairly expensive though, so consult the nutrient deficiency chart to see whether you can't identify the problem first.

Figure 31. Soil probe

Plant tissue tests are taken by removing a sample of one hundred leaves from the tree. Take leaves from the middle of the current growing season's growth. Select randomly, but avoid leaves with very obvious insect damage or residue from spray applications. Again your extension office can process them for you or help you with complete instructions on gathering the sample.

Nutrient Deficiency Symptoms

Nitrogen (N)   Visible on old leaves first. Uniform pale green color. Small leaves, weak shoot growth. Small fruit with good red color.

Phosphorus (P)   New leaves small, bluish green. Purple leaf veins, margins. Soft flesh on stone fruit. Not a big problem for fruit.

Potassium (K)   Visible on old leaves first. Small leaves with curled or rolled edges, especially on peach & pear. Abundant flowers but poor fruit set. Small fruit with poor color.

Calcium (Ca)   Symptoms often visible on fruit first as cork spot or bitter pit. Appears on new leaves first. Yellowed leaf margins. Shoot dieback.

Magnesium (Mg)   Shows up on middle age and old shoots first. Yellowed margins and area between veins. Premature fruit maturity and fruit drop.

Sulphur (S)   Uniform yellowing of leaves. Leaves often take on an orange/red cast. Thin woody shoots.

Boron (B)   Small misshapen young leaves. Abnormal flower development. Fruit and/or bark corking. Wilting & shoot dieback.

Copper (Cu)   Whitish color between leaf veins. Tip withering & leaf fall. Poor fruit set. Small, poor quality fruit.

Iron (Fe)   Seen in tip leaves first. Netted yellowing pattern. Leaf tips & margins die.

Manganese (Mn)   Occurs on older, mid shoot leaves first. Bands of yellow between veins. Paper thin leaves. Poor fruit size & color.

Zinc (Zn)   Leaves on tip of branch stunted. Grow in tight rosette. Irregular, rolled leaf margins. Poor bud break. Reduced fruit set on some but not all branches.

Fertilizer - How Much?

How much fertilizer to apply is something that is difficult to answer with one set formula. It isn't written in stone and can vary with your soil type, amount of soil organic matter, tree age, and volume of crop. Your soil test result recommendation can serve as a starting point if you are preparing a new planting. If you are feeding established trees, the amount of shoot growth can be a helpful guide.

A young, nonbearing fruit tree should start out receiving about ¼ pound of *actual* nitrogen the year after planting. Remember that a fertilizer labeled 33-0-0 is only 33% actual nitrogen per pound of weight so you would need to apply ¾ of a pound of this material to your plant. (1 pound divided by 33%) x (¼ pound required). If the tree has deep green leaves and fifteen to twenty inches of shoot growth annually, continue fertilizer applications at this rate. More than two feet of annual growth means you should reduce the amount of fertilizer provided by ten to twenty percent.

When the tree starts fruiting, reduce the nitrogen level so that annual shoot growth is only fifteen to eighteen inches for peaches, nectarines, apricots and Japanese plums. Shoot growth on spur bearing fruit - apples, pears, European plums - and cherries should average about twelve inches per year. Nutrients other than nitrogen can be applied if deficiency symptoms begin to appear.

Nitrogen fertilizer should be applied in late winter or early spring so that it is released as the tree starts its new shoot growth for the season. During exceptionally cold or rainy springs a small additional application of quick release fertilizer such as calcium nitrate can be made shortly after June drop. Avoid applying fertilizer in late summer and autumn as this stimulates growth and vigorously growing shoots may not harden down sufficiently for winter.

Water - How Much?

If you thought judging fertilizer amounts was tricky, water could give you a real challenge. Soil type and local conditions come into play so much that it is hard to give specific water recommendations. A few guidelines should help you get started. From there you will have to experiment until you find what works for your garden.

When watering fruit trees two things are important to remember. One, avoid letting the tree become so dry that it wilts and two, a deep watering once a week will do more good in developing deep roots than a little sprinkle every few days.

For the young nonbearing tree, a weekly watering of about eight gallons should suffice in all but the driest conditions. A mulch about six inches deep will help preserve moisture too.

Mature trees need about an inch of rain a week. This translates into roughly one gallon of water per square foot of root area. Consider that the root area extends two to three feet out beyond the branches. You can measure rainfall and evaporation by leaving a five gallon bucket out in an open spot and recording the water level weekly. If you are in an area that has a water shortage, watering during fruit expansion is the most critical. This is usually the three to four weeks prior to anticipated harvest.

Mulching

Using mulch in the garden has become a popular way of conserving soil moisture and controlling weeds. Straw, bark chips, gravel, and even specially manufactured plastic mulches suit the purpose well. Leaves and grass clipping work too, but have a tendency to mat down heavily. If you are using any natural plant matter as mulch, you will need to provide extra nitrogen to your tree. Soil organisms that break down these mulches use considerable soil nitrogen in the process. Mulch also provides a nice home for mice, so be sure to rake the mulch back from your tree trunk each fall.

Modifying pH

As mentioned earlier, pH is used as a measure of how acidic or alkaline a soil is. It is indicated on a scale of 0 to 14, with 0 being pure acid and 14 pure alkaline. A measure of 7 is considered neutral. the numbers on the pH scale represent a geometric progression of pH so a soil of pH 5 is ten times more acid than one of pH 6 and one hundred times more acid than a soil of pH 7.

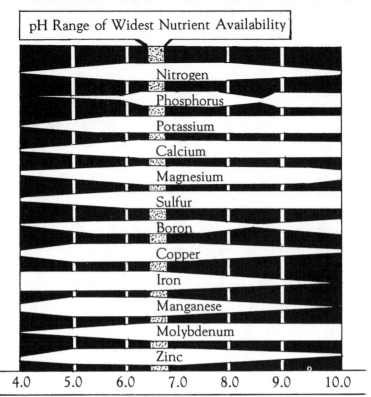

Figure 32. *Nutrient Availability at Different pHs*

Most fruit trees prefer a soil of pH 6.5 to 6.8. In this range, the nutrients required by the tree are most readily available. Soils in wet regions tends to become acid over time due to acids produced by the breakdown of organic matter and fertilizers added to it. Alkaline soils are common in dry climates. If soil tests indicate a need to modify your pH, you can do that by adding calcitic or dolomitic lime to make it more alkaline (raise the pH). Adding sulphur will make it more acid (lower the pH). Both materials are commonly available at garden centers. A general guide to modifying your soil is found below. Specific amounts can be obtained from your extension agent.

Amendment to add to 1000 Ft.² of garden to change pH one point.

Soil type	Lime	Sulphur
Sand	50 lbs.	8 - 10 lbs.
Loam	60 - 70 lbs.	20 - 25 lbs.
Clay	80 lbs.	30 lbs.

126 the Backyard Orchardist

11. Pruning Basics

Pruning fruit trees is one area of horticulture that often seems to hold either mystery or fear for most people, but it really need not be so. Once you understand a few basic principles and follow a step-by-step system, the pruning process and its results really become quite simple and predictable. Also remember that it is hard to harm a tree with a light pruning, but neglecting pruning can eventually lead to a poorly shaped tree that may be too overgrown to produce a good crop. The need for pruning can be looked at from several directions, such as the time period in a tree's life and how pruning affects it or a specific desired effect and how pruning can accomplish it. Let's look first at the various stages of a fruit tree's life and why pruning might be needed at any given stage.

Pruning Young Nonbearing Fruit Trees

Pruning young fruit trees in the years before they start bearing is mostly aimed at helping the tree to establish a strong framework of branches that will be capable of supporting the weight of the future fruit crops. For free standing trees, (those that are not supported by wiring to some type of trellis system), there are three major choices of pruning system. Which one you choose will normally be dictated by the tree's growth habit, or natural way of growing. These three systems are known as the central leader system, the modified leader system, and the open center or vase system.

The Central Leader System. The central leader system is most commonly chosen for trees which naturally grow with one central trunk that has a strong tendency to grow upright. This system is most frequently used for apples. Usually, three or four layers of fairly horizontally growing branches, radiating out from the trunk, are left to grow. They eventually become the major bearing surface for the fruit. These individual, horizontally growing branches are often called scaffold branches because they make up the framework or "scaffold"

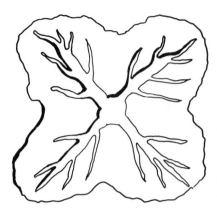

Figure 33. Whorl of branches radiating from the trunk, top down view

of the tree. Each of the layers of scaffold branches normally has three to four branches radiating out from the trunk and spaced fairly equally around it. If you were to look at the ideal central leader tree from the top looking down at the roots, the scaffold branches would radiate out in four compass directions; north, south, east, and west. Each of these groups of branches is often referred to as a whorl. Within each whorl, you may leave six to nine inches between the individual branches. In a tree of ten to twelve feet (which is a typical and manageable height for a fruit tree in the home garden), the first whorl of scaffold branches would occur about two and a half to three feet off the ground. The second whorl would be about three feet above it, with two and a half to three feet between each subsequent layer of scaffold branches. When viewed from the side, a central leader tree will have a cone or Christmas tree shape, wide at the bottom tapering to the top.

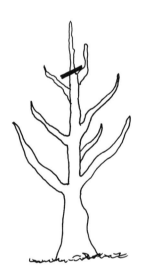

Figure 34. Controlling tree size by heading to a weak leader

Most fruit trees in the home garden are best kept to a maximum of ten to twelve feet in height and three whorls of branches. Once the desired height is reached, the tree is contained by pruning the leader off to a weakly growing leader or by allowing the leader to bend over under the weight of a load of fruit.

Modified Central Leader. As its name implies, the modified central leader system is a variation of the central leader training system. This system is used most often with fruit trees whose initial structure starts with one upright growing trunk and a layer of scaffold branches, much like the central leader. Usually five or six scaffold branches are grown more or less in a continuous whorl spiralling up the trunk. When the scaffolds have reached about six feet from the ground, the central trunk or leader is headed back or cut off to a strong outward growing scaffold branch. The tree is now maintained with an open center (as discussed below) from the point of the heading back cut. This pruning technique is often used to contain a tree at a desired height when its natural tendency is to grow considerably taller. Pears are usually trained to a modified leader system. Tart and sweet cherries are also trained this way. Some upright growing plum varieties and some of the more open growing apple varieties also lend themselves to the modified leader system.

Open Center. The open center training system is used for fruit trees whose natural growth habit is more vase shaped than upright. With the open center system, three to five scaffold branches are developed in a whorl from about three to four feet off the ground. Any leader that may develop above this point is headed back to the whorl of scaffolds. Lateral side branches of the main scaffolds continue to provide additional growth and fruit bearing surface while they grow out and upward in a vase or bowl shape that leaves the center of the tree somewhat open for sunlight to reach inside. This system is most common with peaches and a number of plum varieties that have a naturally open vase shaped growth habit.

In the early years of a fruit tree's life, it is best to prune sparingly. The major goal is to help develop the shape into one of the three just discussed above. Fruit spurs and smaller branches that will be the first to fruit should be allowed to grow. Ideally, the scaffold branches that are left to grow

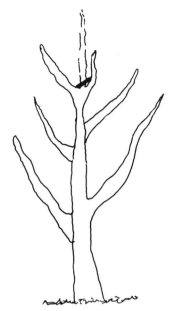

Figure 35. Heading off the leader on a modified leader tree

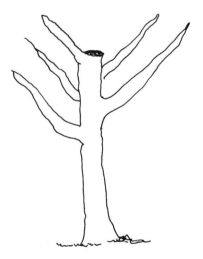

Figure 36. Open center branch arrangement

Pruning Basics   129

should have about a forty five degree angle or greater from where they attach to the tree trunk. Only those branches that do not belong, have poor crotch angles, or compete with the desired branches are removed.

The key to pruning in the tree's early years is to form a desirable branch structure in the beginning so that severe pruning can be avoided later. Your goal in the first three years should be:

Year 1 - Prune as lightly as possible to form scaffolds

Year 2 - Encourage balanced growth of lateral and secondary scaffold branches

Year 3 and beyond - Keep the tree canopy open to sunlight and encourage even distribution of fruiting wood

Pruning the Bearing Fruit Tree

Once the major scaffolds of the fruit tree have been developed and it starts bearing fruit, you will want to shift the focus of your pruning somewhat. Now the goal will be to maintain the established tree shape and height while allowing enough sunlight into the interior of the tree. Occasionally, you will also need to remove broken or diseased branches and open up some access holes so that you can get inside the tree canopy to pick all the fruit.

Sunlight is a necessary element for photosynthesis and fruit bud development. For this reason much of the pruning that is done on "mature" bearing trees is done to allow sunlight into the tree. Another reason for pruning the bearing tree is to stimulate a modest amount of new vegetative growth to replace older wood whose fruit production is decreasing. Which of these effects is desired will dictate at what time of the season you will want to do your main pruning.

If you are pruning primarily to open the tree up to allow sunlight in, you will want to prune sometime during the early to middle part of the active growing season on all but the most fireblight susceptible varieties. This will normally be from May through August, varying somewhat with the growing season in your area. You may hear this referred to as "summer pruning". By pruning once growth has started, you avoid some of the growth of excess "water sprouts" or suckers that often occurs if the tree is pruned during the dormant season.

When your main goal is to invigorate the tree and encourage new vegetative growth, you should prune during the dormant period. Typically the winter months anywhere from November to March or

April, depending on your location, are a suitable time. Trees should have shed their leaves, active shoot growth should be finished for the season, and green tissue should be sufficiently hardened down before starting dormant pruning. Otherwise, since pruning stimulates growth, the tree may be prompted to grow again, leaving tissue susceptible to winter cold injury. More hardy trees such as apple or pear can be pruned almost any time during this dormant period. More cold sensitive trees such as cherries or plums are best pruned closer to the end of winter so that they may begin growing again soon and avoid cold injury and drying of the pruning cuts. Peaches are often pruned just after they bloom, especially in cold climates.

When pruning the mature bearing tree, it is often easiest to have a simple system in mind. When selecting branches to prune think of removing them in this order:

1) Cracked, broken or diseased branches.

2) Low hanging branches that drag on the ground or hang below horizontal should be cut out. The same goes for watersprouts and double leaders.

3) If two branches are rubbing on each other, remove the least needed one.

4) Where two scaffold branches are crowding each other, remove one. Generally this should be the lower one, but common sense will tell you if one is noticeably better than the other. Your goal is to allow about four feet vertically between parallel branches.

5) Lastly, create holes for sunlight by cutting out several large lateral branches in the main canopy of the tree. Removing them will give you holes that also allow you to reach inside the tree canopy and pick the fruit.

6) On a central leader tree, you may also want to cut back the higher scaffolds by 20 to 50% of their length to maintain the Christmas tree shape.

If these steps are kept in mind, pruning a fruit tree can be as simple as 1-2-3. For most people the first cut is the hardest, but with a little bit of practice, you'll be moving right along. Common sense and a little bit of experience will show you which branches to prune. A regular annual pruning will make the job easiest and will also be best for the tree, but don't worry too much because many fruit trees get along fine being pruned only once every two or three years.

Rejuvenating the Neglected Overgrown Tree

Most questions on rejuvenating an old, overgrown tree come from people who have acquired a "gentleman's farm" in the country. Often these are old farmsteads that include a neglected fruit tree or two. The trees are likely to be apples or pears, since they are the most inclined to survive in spite of neglect.

With some patience and hard work it is well possible to rejuvenate the forgotten tree. Several factors should be evaluated first, however:

1) Is the tree still reasonably alive? Assuming it is rather large, if the main trunk and at least one third of the major branches are still healthy, the possibility of rejuvenation exists.

2) If available, sample some fruit. Your personal taste will have to serve as the judge of whether it is indeed worthy of many hours of sweat and attention.

3) Ask yourself why you are saving this tree? Is it to preserve something special or is it "just because it's there". You will need to put more time and effort into slowly bringing an old tree back than will be required of a newly planted tree. The actual long term cost of a newly planted tree is no more and often less than the care of a large old tree. Also, if your major goal is to preserve a variety you fear will otherwise be lost, you could gather scion wood from the old tree and propagate new trees with the normal grafting procedures.

All that said and done, if you have decided you want to pursue renovating the old tree forge ahead! Remember that there is no need to be afraid that your pruning will do great damage to the tree. Those first cuts are always scary, but they will be a benefit to the tree if it is not already on its last leg. Should your renovation be unsuccessful, it is probable that the tree was already beyond hope, not so much that you did anything incorrect.

Begin renovating by cutting out all obvious dead wood. (This could involve using a chain saw.) When cutting large limbs, remember to use the techniques illustrated in Figures 37 and 38. Of the live wood that remains, one third to one half can safely be removed. Concentrate on those limbs whose structure is the poorest. That is, limbs growing across or rubbing on others, limbs growing too low to the ground and limbs growing straight up.

Continue this process each year until you have obtained a desirable tree shape. You may find winter the easiest time to prune. The leaves are not on the tree and your view of the branches will be clearest then. Annually during mid summer, remove the abundance of suckers that will appear at many of your large pruning cuts. If done when the suckers are eight to twelve inches long, they can often just be bend downward by hand and will pop off easily. Continue to systematically prune the tree annually for three to four years. You can remove one third to one half of the over growth each season. In several years you should have a manageable tree. From that point on prune it as you would any mature tree of its type.

One major caution to observe: By removing so much wood you may fool the tree into thinking that it is being well fed, now that the roots have fewer branches to feed. The tree may respond by growing very lush, soft green shoots. Especially in pear and apple, these shoots can be very susceptible to fireblight. Do not encourage additional shoot growth by also feeding the tree large doses of fertilizer. Feed only a small dose prudently in the first year or two after major renovation. Later you can increase the dose based on the tree's growth.

Proper Pruning Cuts

Most people who are concerned about pruning a valued tree or shrub are afraid of damaging the plant beyond repair. In most cases this is unlikely to occur, but proper attention to how the cut is made can help the tree heal faster after pruning.

In pruning, you will likely be making one of two types of cuts. You will either be cutting shoots and young branches with diameters of less than an inch or you will be cutting more substantial size wood. Cuts on most small diameter wood can be made with hand clippers, limb loppers, or a small hand saw. Larger wood will probably need to be cut with a saw of some sort.

The basic point to remember in all pruning is that the cut should be as close to the collar of the branch as possible without cutting into it and, as much as possible, avoid leaving branch stubs sticking out. If you look closely where a branch grows out of the trunk, you will notice an area at the base of the branch that is somewhat thicker than the branch itself. This area is the collar. It contains a natural chemical barrier that stops decay and has cells that will rapidly grow scar tissue to heal the area where the branch is removed. The scar tissue will heal most easily when it has a clean

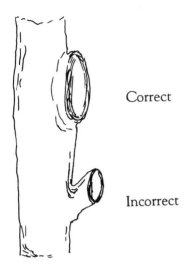

Correct

Incorrect

Figure 37. Pruning cut - right & wrong

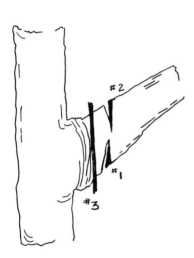

Figure 38. Steps in cutting a large branch

smooth surface to grow over; consequently it is important to try to make clean, flush pruning cuts that do not leave stubs behind.

Making the nice smooth cuts mentioned above is easy. Just work with good sharp tools. For small cuts, one firm clip of the shears should do it. If you are cutting large heavy branches, though, the weight of the limb can sometimes cause the bark to tear, as a branch breaks before being fully cut. To avoid this unnecessary injury to your tree, large branches should be cut with a three part cut. First make a cut on the bottom side of the branch to be removed. The cut should be about two to three inches away from the collar and about one third of the diameter into the branch (or until the weight of the branch starts to pinch your saw). Then from the top of the branch, complete the cut. The branch should cut away clean in most cases, but if it does not, any bark tearing will most likely stop at the cut you made on the bottom side of the branch and avoid tearing a big strip of bark down the trunk of the tree. With a third cut, remove the remaining stub.

Most pruning cuts fall into one of two types, thinning out cuts or heading back cuts. Thinning out cuts are what their name implies. They are used to remove unwanted branches and are made by cutting a branch back to its base. In the case of a scaffold branch, this would mean removing it back to the tree trunk. Lateral branches are thinned out by cutting them back to the scaffold branch from which they originated. Thinning out cuts usually have the effect of encouraging other neighboring buds to begin growing.

Heading back cuts involve cutting the tips of branches back a given length, not necessarily to a main trunk or scaffold. Heading back cuts are commonly used to contain a tree that may be growing too large for its allotted space. A heading back cut tends to temporarily stop further growth of the branch that was cut back and has little effect on the growth of other neighboring buds or branches. In most cases it is preferable to thin out rather that head back a branch.

Pruning Tools

Many jobs are most easily done when using the right tools. Pruning is no exception. So let's look at some of the tools available and how they are best used.

Pruning shears - These will be one of the first pruning tools you will use, as they are compact and handy for trimming the fruit tree in the first years of its life. Pruning shears come in two basic types, by-pass shears and anvil head cutters.

Anvil head shears cut by pressing the sharp blade against the flat anvil portion of the head. Anvil cutters have a tendency to crush the branch tissue but are able to cut larger stock than a by-pass shear. With by-pass shears the cutting blade works like a scissor. The general rule for by-pass shears is they will cut wood up to the diameter of a thumb. (Do be careful to keep those thumbs and fingers out of the way as a sharp bypass shear

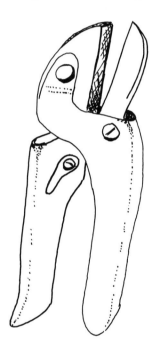

Figure 39. Anvil head pruning shears

won't discriminate!) By-pass shears will cut more cleanly than an anvil shear, but are more likely to be damaged if you try to cut something they aren't meant to cut. If you are going to buy only one pruning shear, your best bet might be a high quality by-pass shear.

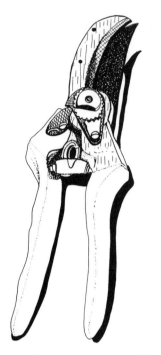

Then, take care of it as you would with any treasured tool and don't allow it to be used for anything other than its intended purpose. It will serve you well for many years.

There are many styles and sizes of hand grips for pruning shears. Look at several. Try them out and see what feels best in your hand. If you are left handed, there are a few styles made for you, too, and it is worth investing in one to avoid hand fatigue if you are doing a lot of pruning.

Also consider a pruning shear that allows you to remove the blade and sharpen it periodically or replace it if necessary. This will help both in keeping your pruning cuts clean and in not tiring your hand when doing much pruning.

Figure 40. *By-pass pruning shears*

Limb loppers - For cutting branches larger than what pruning shears can handle, limb loppers are the next step up. Loppers normally have handles twelve to thirty

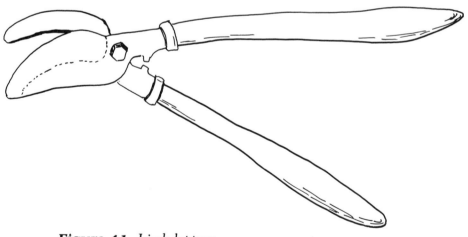

Figure 41. *Limb loppers*

136 the Backyard Orchardist

inches long to give more leverage when cutting branches up to an inch and a half in diameter.

Loppers are used by holding them with both hands when cutting. The cutting heads are often similar in style to pruning shears, being of either the anvil head or by-pass type. Choose your lopper for the same reasons as you would shears; they should be sharp, sturdy and comfortable in your hands. In choosing loppers be sure that the handles are good and solid, since they can be under considerable stress when cutting thick branches and you do not want them to give way suddenly. Also, if you are doing much winter or dormant pruning, metal handles may tend to be colder than wood.

Pruning saws - Pruning saws come in a number of sizes and shapes, but all do essentially the same work. They are most suitable for cutting medium to large wood that loppers can't handle or reach. A pruning saw may be a small blade that folds into the handle, a bow type saw with a thin blade for tricky places, or a pole saw on a long handle to reach the tops of tall trees.

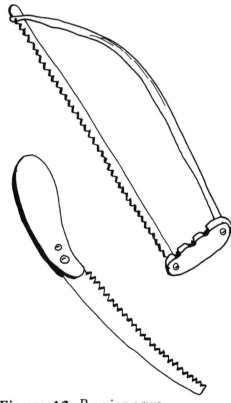

Most pruning saw blades have coarse teeth that are offset with teeth alternately bending to the right and left down the saw blade. The teeth are often sharpened in only one direction so that they cut while being pulled, but do not bind up when being pushed back for the next stroke. Due to the offset teeth most blades will make a fairly wide cut which helps shed wet sticky sawdust from their path. Many of the blades themselves are narrow or curved on the end in order to reach into tight places between branches. Most pruning saw blades can be removed and sharpened, for just a few dollars when dull or replaced when worn out.

In choosing a saw consider

Figure 42. Pruning saws

what type of pruning you will be doing most. If you have all dwarf trees that you can reach from the ground, if your space is very compact, or your tree is planted close to a building, a hand saw will be most convenient. A pole saw is the best choice if you need to reach the tops of larger trees. Once you get used to them, pole saws are very handy because you can stand back from the tree and get a complete overview of the tree while you prune. Many commercial fruit growers do all the pruning of their mature trees with pole saws as they are fast, light weight, convenient and relatively inexpensive. If you are doing much dormant pruning, consider a saw that you can hold comfortably with your work gloves or mittens.

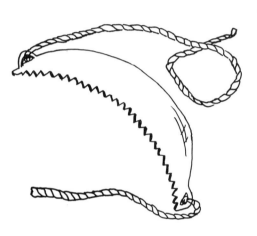

One word of caution on saws. The one saw to avoid is the type with a string attached to each end of the blade and meant to be thrown over an out of reach branch. (In most home gardens your trees will not be so large that you would even think about needing to use this type of saw.) Since there is no way to make a bottom cut with these, they will often cause considerable bark tearing if cutting a big branch. If safely possible, climb into the tree and use a hand saw or use a pole saw for those hard to reach spots. For smaller diameter wood, the next tool may do the trick for you.

Figure 43. A pruning saw to avoid

Pole Pruners - A pole pruner is somewhat of a hybrid instrument, a cross between a pruning shear and a pole saw. Normally the cutting assembly consists of a spring loaded by-pass blade that is operated by pulling a string or metal rod attached to it. The whole assembly is attached to a pole, much like a pole saw. A handle around the pole often secures the other end of the string or metal rod that operates the cutting blade. A pole pruner can be very handy if you have small diameter wood that is too far away to climb to safely, but small enough to cut with a snip of the pruner. If you have large, vigorous growing pear, sweet cherry or apricot trees, consider investing in a pole pruner.

As with most tools that you expect to use skillfully and often, it is wise to invest in good pruning tools even if the better ones cost a little more. They will pay you back many times over in the way by making pruning easier. Do remember, too, that they will serve you just as well on the other shrubs and trees in your yard as they will on your fruit trees, so you will probably use them more than you originally anticipated.

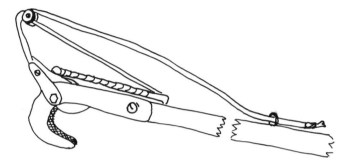

Figure 44. *Pole pruner*

12. Growing Fruit Trees in Containers

Even as an apartment or condominium dweller you can enjoy the benefits of growing fruit trees. All that is needed is the selection of the proper dwarf fruit tree grown in a suitable container. Container growing is also useful in northern climates where you may want to bring a cold tender tree inside or in hot dessert climates where you need to protect your tree from excessive heat and sunburn. If you thought you might have to forgo your dream of enjoying your own tree ripened fruit due to a challenging growing climate or limited space, container growing can put your dream within reach.

Genetic Dwarf Trees

In recent years, several plant breeding programs, both by major universities and private individuals, have produced a number of fruit tree varieties that can be successfully adapted to container growing. Some of these trees are small in size due to the use of dwarfing rootstocks and spur-type scion varieties. Others are true genetic dwarfs. More information on dwarfing rootstocks and spur-type varieties can be found in Chapters 4 through 9. For the greatest success and least effort, you would do best to select a fruit variety that is suitably hardy for your climate whether grown in a tub or not.

Genetic dwarfs are trees that have been selected for their miniature size after years of breeding. Not only are they short in height, most reaching only four to five feet, but their branches are also compact growing, so they often take on the appearance of a round shrub growing on a single trunk. One thing that is not miniature about the genetic dwarf trees however is their fruit. Identical in size to that of their full size cousins, it is also similar in taste.

A relatively extensive selection of genetic dwarf peaches and nectarines is available. Although none of them is truly winter hardy, they can be moved inside for protection in the coldest climates. (See the climate zone map on page 25 to determine chilling requirements in your area.) Genetic dwarf peach varieties that are well adapted to container culture include:

Low chill varieties - Southern Rose and Southern Flame.

Moderate chill varieties - Bonanza, Garden Gold, Garden Sun.

For colder areas - Honey Babe, Sunburst, Compact Redhaven.

Suitable nectarines include:

Low chill varieties - Garden Beauty, Garden Delight, and Garden King.

Moderate chill - Sunbonnet.

For colder areas and high chill - Golden Prolific.

With the exception of Compact Redhaven (which is not a true genetic dwarf), most of these genetic dwarf trees can be contained to a size of about five feet high when container grown.

Most of the other fruits have a more limited choice of true genetic dwarfs to choose from. The pear variety Little Princess is a natural dwarf and some self-fruitful scion varieties compatible with Quince rootstock can be dwarfed sufficiently to grow in containers. The tart cherry Meteor has a spur-type growth habit and is considered by many to be a semidwarf tree. North Star tart cherry is the only true dwarf choice. Sometimes the sweet cherry variety Garden Bing is sold as a dwarf tree. Experience has shown, however, that branches of this variety often revert to full size over time. It is therefore an unreliable variety for container growing. Apple trees suitable for container growing include spur-type scion varieties grown on Malling 27, 9, and 26, or P-22 rootstock. Apricots offer a choice of Aprigold and Garden Annie. Pixy, the most dwarfing rootstock for plum, is useful for container culture, if you can obtain it.

Choosing a Container

The container for your dwarf tree should have several important characteristics. It should:

1) be non toxic
2) have proper drainage
3) be two to three inches larger than the root ball of the tree
4) be movable, if possible.

In choosing your container, look for something made of clay, metal or untreated wood. Cedar is a good choice of wood as it naturally resists rot brought on by exposure to the elements. Oak half whiskey barrels work well too. Beware of dyes in ceramics or pentachlorophenol wood preservative. They may leach into the soil and be toxic to your tree.

When starting out with a one year old bare root tree, a five gallon can or bucket is quite suitable. Over the next two or three growing seasons, the tree will have to be repotted to successively larger containers, ending up in something the size of a bushel basket or half whiskey barrel. The final planting container should be at least eighteen inches across and equally as deep. You can choose an attractive container as your planter. A basket or specially built wooden box can serve as a decorative surround, too.

Make sure the container you choose has adequate holes for moisture drainage. If not, be sure to put a layer of several inches of coarse gravel in the bottom so that roots do not sit in continually saturated soil.

Remember that your tree will need to be root pruned and repotted every two or three years. Keep this in mind in your selection. A planter that can be disassembled will make the job easier. If your tree will need to be moved seasonally to protect it from the weather, consider a container that is light weight or place it on a small platform with wheels to simplify relocation.

The Soil Mix

The container grown tree is totally reliant on the small amount of soil available for all its nourishment. Therefore, it is important that the soil mix be properly prepared to provide what the tree needs. The soil will need to hold moisture adequately and yet drain excess water at the same time. The roots need some access to air and of course nutrients must be provided. The soil should also be free of weed seeds, diseases, and insects.

In selecting the soil, you have two choices: You can buy a commercial potting mix (this is easiest if you are only potting one or two trees); for larger numbers of trees, you may find it economical to prepare your own mix.

Most commercial mixes are made up of equal parts vermiculite for drainage and peat moss for moisture retention. Supplemental nutrients for initial growth are also included in most commercial mixes. You can make your own soil mix using one of the following combinations:

1) one part fine sand and two parts finely ground bark.

2) equal parts peat moss, ground bark, and fine sand.

3) equal parts perlite, vermiculite and peat moss.

4) two parts peat moss, one part perlite and one part fine sand.

To one cubic yard of any of the above mixes add a fertilizer mix of five pounds 5-10-10 commercial fertilizer and five to seven pounds ground limestone. To properly make up your soil mix, start with a separate pile of each ingredient. Blend them together by shoveling them onto one cone shaped pile. Add each shovelful to the top of the cone so that the ingredients can mix as they roll down the pile. When you have one completed pile, build a second pile to allow additional mixing. Continue this two or three more times until fully blended. Dampen the mix as you go. Using warm water will make it much easier to wet the peat moss. Finally, mix in the fertilizer with the same technique. Your mix is now ready for use. It can also be stored in a plastic bag or trash can for later use.

Planting and Repotting

Planting the fruit tree in a container will be similar to planting it in your garden. Prepare the pot by placing broken pieces of crockery or screen over the drainage holes so that soil does not flow out during watering. Then place a shallow layer of soil mix in the bottom of the pot. Position your tree so that the location of the graft union relative to the soil line will be like a garden planted tree (for details see page 40). When full, the soil line in the pot should be about a quarter inch below the rim of your container. This will allow for settling when you water.

Figure 45. Properly potted tree

As they grow in a container, feeder roots tend to cluster around the outside of the root ball at the wall of the container. This reduces the area available to take up nutrients. To remedy the problem, you will need to occasionally repot the tree. In the first three to four years, you will probably repot the tree annually into a larger pot. After this, repot as needed. Your tree will indicate its needs by growing more slowly. When possible, time your repotting for early spring as the tree is about to enter a new growing season.

To repot, remove the tree from its pot. With a sharp knife cut away about one inch of roots on the side and bottom of the root ball. Place a fresh layer of soil mix in the bottom of the pot. Gently break the root mass apart a bit with a fork or your fingers. Reposition the tree and fill soil mix in around the sides of the pot. Water thoroughly and add more soil where settling occurs. Prune the top growth to balance with the amount of roots removed. Healthy new root and shoot growth should soon follow.

Feeding and Watering

How often to water your tree will depend on the type of container it is in, as well as the natural rainfall and evaporation. Never allow your tree to wilt. Continuously soggy soil can be harmful, though, too. Check the soil moisture regularly by digging down an inch or two and feeling the soil with your fingers. You may need to water daily if the weather is warm and windy and you have used a clay pot. On the other hand, a tree in a plastic pot may need only a weekly watering in cool temperatures. One way to preserve moisture and keep roots cooler is to use a two inch thick layer of coarse bark mulch in the top of your planters.

When you water, do it gently but thoroughly. Your tree will benefit more from a twice weekly soaking than a daily superficial sprinkle.

How often to feed your containerized fruit tree will be dictated by how often it is watered. Frequent watering will cause the soluble nutrients to leach down quickly through the growing medium. This is especially true with synthetic commercial soil mixes. Let your tree be a guide in how often to fertilize. If it shows yellowing of the leaves or other obvious nutrient deficiency symptoms, increase the frequency of your fertilizer applications. Compared to a garden grown tree, the container planting has a smaller root area. Therefore, it will need more frequent fertilizing. A good rule of thumb is to feed half the recommended dose of a complete fertilizer every two weeks. Liquid fertilizers are easiest to mix accurately and are also less likely to burn roots. Continue fertilizing from early spring until mid July.

At this time, stop fertilizing so that the tree can slow its growth and toughen its tissue before the dormant season.

Mineral salts from fertilizers and hard water frequently build up in container grown plants. This can cause leaves to brown around the edges. To avoid this problem, purposely flush the pots every two or three months. Gently run water into the pot for fifteen to twenty minutes so that it slowly flows out the drainage holes. Do not use softened water for your trees as it contains a high amount of salt and will only increase this problem.

Keeping the Container Grown Tree Healthy

Many of the activities of keeping a fruit tree healthy apply to a container grown tree just as they do to a garden planted one. Along with repotting which was discussed earlier, modest pruning of the branches is also beneficial. Procedures are similar to those outlined in Chapter 11. The amount of wood removed would be reduced however, in keeping with the container grown tree's size. Buds can also be pinched in the summer to further control the size of container grown trees. In this procedure, the tips of vigorous growing branches are simply pinched out by hand, much the way many house plants are occasionally "pinched". The effect is much like a heading back pruning cut.

Insect and disease control methods are also the same as for full size trees, although more careful attention should be given to control of mites.

Winterizing the Container Planting

As recommended earlier in this chapter, it is best to choose a container grown plant that will be winter hardy for your growing area. If you have not done so, the tree will need to be brought inside during the winter. Remember though, that it needs to fulfill its chilling requirement during the dormant season. Keep it in the coolest possible location, but protected from damagingly low temperatures - an unheated basement would work.

If you have selected a winter hardy variety, it can be kept outside year around. Because your tree is in a container, you will have to make some adjustments though. Container grown plants are often more vulnerable to cold than their garden grown relatives. While garden grown trees' roots receive some insulation from the large, below ground soil mass, the soil in a container may be subject to repeated quick freezes and thaws. Not only is this stressful on the roots, it can cause the tree to heave out of the soil in the pot and break roots as well. Fortunately, if your tree is otherwise hardy for your winter climate, there are steps you can take to protect it.

Once the tree has hardened down in the fall and shed most of its leaves, you can begin to build its winter shelter. Prune out broken or dead branches as well as any excessively long ones. Then firmly tie the bearing surface up with twine, pulling the branches together. Water the soil well and mulch with several inches of straw.

Next, fashion a cage of chicken wire or comparable garden fencing. Make a cylinder about twelve inches wider in diameter than the tree canopy. Staple or tie both ends of the fencing to a vertical wood slat to fasten the cylinder closed.

Gently fill the cage with loosely fitting, dry leaves or straw. For a tall tree build a second tier on your cage to reach the full height of the tree. When the cage is fully stuffed, surround it with burlap and cover it with plastic or water proof roofing paper. Tie the cover in place for the winter.

As spring arrives, gradually remove your protective layers,

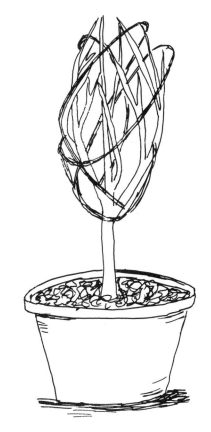

Figure 46. Tieing up branches in preparation for winterizing

allowing the tree to reacclimate over a week or two. Save your materials to reuse next winter.

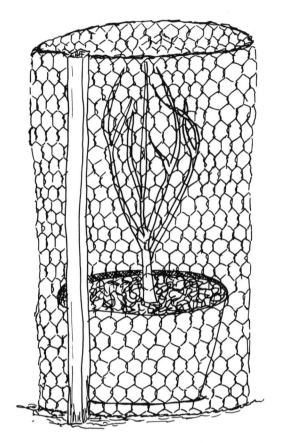

Figure 47. *Preparing a winter shelter*

13. Flowering, Fruiting and Thinning

One of the basic concepts that you should understand when growing fruit trees is how the tree, flower and fruit develop to eventually yield a crop. Many of the things you do in caring for your tree, from fertilizing to pruning, can effect these processes. Since some of this development takes place inside the tree and buds, it is not visible to the naked eye, yet it is extremely important to determining the success of your crop.

In the first years of growth, a fruit tree is in a juvenile state during which time all of its energy goes to producing new shoots and roots. Often this time period is referred to as the vegetative phase, since the tree is producing only vegetation. This phase may last anywhere from a year or two with some of the dwarf apples to seven or eight years with seedling apples, cherries, or pears. How long this stage of the tree's growth lasts can be somewhat influenced by the rootstock that the tree is grafted to or by several mechanical techniques. In technical literature, a particular rootstock may be referred to as precocious. This means that this rootstock will cause the tree to move out of its juvenile stage at an early age. It is not known exactly how or why a tree starts to move out of a purely vegetative stage and into a more mature stage where flowering and fruiting is possible, but gradually this does occur. Generally the lower parts and the interior of the tree (that part closer to the main trunk) will keep its vegetative character and continue to develop new shoot growth.

Fruiting wood is generally seen on the ends of the shoot tips or outer shell of the tree. This tendency can often challenge you in trying to get maximum potential production from a tree, since much of the interior of the tree has the potential to produce fruit, too. Certain pruning techniques can be used to counterbalance this tendency somewhat. Fruit producing buds can also be found on small modified branches called spurs. The spurs are often found on the inside of a well pruned tree and can be responsible for much of the fruit bearing surface of apples and pears.

Understanding Fruit Bud Development

Another fundamental concept to understand is how and when the fruit buds are actually formed within the tree's tissue each season. Once a fruit tree has reached a stage of maturity in which it will fruit, it has started on a regular path of developing many flower buds each season that will eventually result in a fruit crop. (You may see the terms fruit bud or flower bud used in literature. The two terms are synonymous and used interchangeably.) This shift from producing purely vegetative buds to producing both vegetative and fruiting buds is normally taken as the sign that the juvenile stage has ended.

Most people do not realize, however, that the formation of the fruit bud is actually taking place the season before the flower becomes a visible bloom on the tree. In early summer (usually mid June in the upper midwest or very close to the period of June drop on already bearing trees) the bud tissue begins to differentiate and form the beginning of a flower bud. Many of the sexual parts of the flower are formed before the tree goes into its winter dormant or rest stage.

While in the dormant stage, internal growth processes continue at a very slow pace. During this time the fruit tree needs to be subjected to a certain amount of time below forty five degrees or fruit and leaf buds may not develop and open normally. Although rarely a problem in northern climates, varieties with special low chilling requirements need to be selected in some of the mild growing regions of the south and the west coast. Recent research has shown that an interruption of the chilling period can occur if winter temperatures climb above sixty degrees. Under these conditions, degree units are actually subtracted from the chilling effect. On occasion a tree planted near a stone patio with a south facing sun exposure may encounter these conditions and fail to form buds properly. In the spring, the tree may be slow to leaf out or have malformed flower buds. Consider the likelihood of this occurring in your locale when choosing your fruit tree's location. Typical chilling requirements can be found in Table 24.

Table 24 - Typical Chilling Requirements

Apple	1,000-1,500 hours
Apple, low chill	600-800 hours
Apple, sub tropical	200-500 hours
Apricot	700-1,000 hours
Cherry, sweet	1,100-1300
Cherry, tart	1,200
Peach & Nectarine	500-1,200
Peach, low chill	200-400
Pear, domestic	1,200-1,500
Pear, Asian	900-1000
Plum, domestic	800-1,200
Plum, Japanese	700-1000

Following the required chilling period, flower bud development continues with the final formation of pollen and the ovules in early March, shortly before the blossom opens.

Managing Growth and Fruiting in the Early Years

Using the understanding that you now have, consider how you might manage a fruit tree during its early years. In the first years after planting, your primary goal should be to help the tree develop a strong branch structure that will be capable of supporting the weight of the crop in later years. Sometimes a very precocious tree starts producing fruit in its first year or two. As much as you may hate to, you should remove the fruit. At this stage the root system may not yet be able to meet the nutrient needs of the growing fruit and may try to do so at the expense of shoot growth. The weight of the fruit may also bend the young supple branches into positions that could jeopardize future fruiting capacity. The best things to do for the tree are to provide it with adequate (but not excess) water and nutrients as recommended in Chapter 10. Your goal is to encourage about twenty four inches per year of strong growth on each new shoot.

Just as the level of maturity of a fruit tree has an influence on its fruit producing capabilities, the degree of horizontal orientation of a branch can also effect its fruiting ability. When a branch grows very upright, as rapid growing juvenile growth tends to do, it has a tendency not to fruit. If you encourage that same branch to grow more horizontally, ideally at about a forty five degree angle from vertical, hormonal influences take effect that encourage the branch to produce more fruit buds. A wider angled branch is also stronger and more capable of supporting the resulting fruit. There are several ways that the home orchardist can encourage the new branches to grow at a proper angle to the trunk.

To develop the proper angle the branches can be gently spread and held apart during the growing season. This operation is best done when the branch has put on sufficient growth of twelve to eighteen inches but the shoot is still soft and flexible. In the midwest and upper east coast, proper growth has usually been achieved by mid to late June (this may vary some if you have an earlier or later growing season). On the west coast and in the south, where spring starts earlier, you may be able to spread the branches a few weeks earlier. In the first season or two, gently bend the branch out from the trunk about forty five degrees. Be careful not to bend it too far, which may crack the new branch off. With apple and tart cherry trees, insert a round wooden toothpick or a specially designed plastic spreader between the branch and the trunk. Then gently press on the branch just under the toothpick to be sure that it will hold securely between the trunk and branch.

The same bending procedure can be used for peaches and sweet cherries. Since these types of fruit trees tend to ooze and not heal well when pierced by the toothpick, a spring type clothespin can be clipped and wedged in to maintain the angle instead. When using the clothespins, check them once or twice in the later part of the summer to see that they are not constricting the

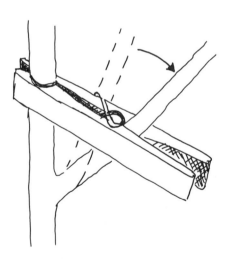

Figure 48. Using a clothes pin to spread peach or sweet cherry branches

branch. During the late summer and early fall, branch growth will normally stiffen and increase in diameter rather than continue to gain length. If the clothespin is too constricting, you can often remove it once the branch has stiffened into the desired angle.

Pears, plums, and certain apple varieties (i.e. Cortland, Northern Spy, and Rome) have young growth that can be quite willowy. It is long, thin, and very flexible. This makes it difficult to hold in place with a spreader. Another method of spreading the branches to a wide angle is useful for these varieties. Using plastic baling twine or other string that will not decay rapidly, the branches are tied down and the end of the string secured in the ground with a small metal "W" clip sold for this purpose. Check periodically that the string is not constricting the growth of the branch as it increases in size.

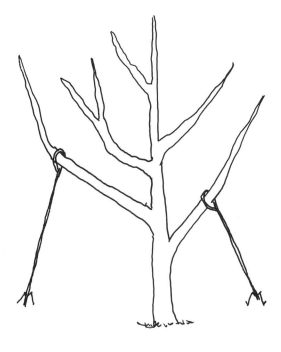

Figure 49. Spreading branches by tieing

Once the tree has a sufficiently developed scaffold of branches, you can move on to encourage the development of fruit if the tree is not already showing signs of producing a crop. Each of the different fruits has its own natural timetable for starting to fruit, but you can generally expect it to be as outlined in Table 25. The type of rootstock the tree is growing on, the variety, and the care the tree has received can all effect this timing. Consider these as guidelines and consult the chapters on the specific fruit for variations.

Table 25 - First Bearing Age

Apple, standard	8 -10 years
Apple, dwarf	3 - 6 years
Pear	8 - 10 years
Sweet cherry	5 years
Tart cherry	4 years
Apricot	4 years
Plum	5 years
Peach/Nectarine	4 years

Flowering, Pollination and Fruit Set

When your fruit tree is several years old and flowering nicely, it is taking its first steps toward a productive harvest. To understand some of the other processes that must occur before you can harvest that prized fruit, it helps to understand some basic biology.

Fruit trees have what are known as complete flowers. That is they have both the male and female reproductive structures contained within each blossom. The accompanying illustration shows the typical arrangement of a fruit blossom. The female parts are:

The stigma - the sticky receptive surface that pollen becomes attached to during the pollination process.

The style - the "stalk" that the stigma is located on and through which the germinated pollen grain grows until it reaches the embryo sac.

The ovary - the fleshy area that serves to protect the ovule and also eventually becomes the fleshy part of the fruit.

The ovule - the structure inside the flower that contains the female genetic material and that will eventually become the seed or pit of the fruit.

The stigma, style and ovary together are often called the pistil. The male parts of the flower are:

The pollen grain - a small dust size particle that serves as the location of the male genetic material.

The anther - the structure that houses two sacs that contain the many pollen grains.

The filament - a stalk upon which the anther is held to elevate it and make the pollen more accessible.

154 the Backyard Orchardist

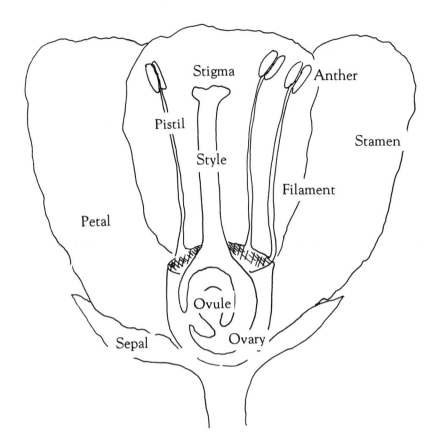

Figure 50. *Parts of the fruit blossom*

Together the anther and filament are often called the stamen.
The petals and sepals serve to protect the reproductive parts and
the colorful petals may also serve as a visual way to attract bees
to the flower.

A beautifully blooming fruit tree does not guarantee that you will
have a crop to harvest, however. In order for successful fruit development to occur, several processes must take place between the male
and female part of the flower. Assuming that the pollen has
developed properly, it must then be transferred from the male anther
to the female stigma. This process is known as pollination and is
usually aided by a pollenizer, most often a bee. In plant breeding

Flowering, Fruiting and Thinning   155

operations this process can be selectively carried out with a small paint brush. After pollination the pollen grain must germinate and grow down the stigma to the ovule where the male genetic material from the pollen tube is united with the female genetic material in the embryo sac. This uniting process is called fertilization. Most of this process will not be visible to the human eye, but shortly after bloom you will notice that the flower petals have fallen off and the fruit presumably begins to develop. Not all blossoms will have been fertilized and shortly after bloom a number of flowers will often drop from the tree and never develop. This is perfectly normal and due to the fact that no viable seed was formed in these fruit. If the pollination and fertilization processes have been successful, the fruit will continue to grow and eventually you will have a mature edible fruit. If weather conditions have been cold or damp or bee activity has not been very heavy, these two processes may not take place properly.

June Drop

Often the home orchardist becomes puzzled when the fruit appears to be developing and then, suddenly, from a week to several weeks after bloom a sizeable number of fruits start dropping to the ground. When drop occurs several weeks after bloom and the fruit has already started to develop to about ¼ to ½ inch diameter, the drop is referred to as "June drop" (although it may occur as early as May or as late as July, depending on what growing region you are in). Alarming as it may seem to watch a large amount of young fruit drop from the tree, the results are not usually as dismal as they appear at first glance. Be aware that only as little as 5% to 10% of a fruit tree's bloom actually needs to develop into mature fruit for a full crop to result. The June drop is nature's way of eliminating potentially inferior or excess fruit. Mother Nature is quite sophisticated in knowing her limitations. Often in years when bloom is very heavy, the June drop is also very heavy: this protects the tree from having to feed or carry more fruit weight than the branches can reasonably support.

Biennial Bearing

At this point, mention should be made of biennial bearing, a problem that occurs with some fruit varieties. Trees that experience this condition tend to bear fruit only every other year. One year the crop is very heavy, requiring most of the tree's energy. Consequently, few new fruit buds are formed and the following year the tree has little bloom and limited or no crop. This problem is most often seen in apples, but can occur in other fruit as well. Some varieties that have a reputation for biennial bearing include Baldwin, Cortland, and Yellow Transparent. Fortunately, in most cases, several things can be done to control biennial bearing. Annual pruning will help control the number of excess fruit buds on the tree. Removing excess fruit from the tree soon after bloom is also helpful; which brings us to the next section.

Hand Thinning

Nature will often shed the obvious excess or frost damaged fruit for the tree's well-being. However, the fruit grower may want to do some additional hand thinning for the purpose of:

1) Encouraging additional size in the remaining fruit
2) Removing insect damaged fruit
3) Improving fruit color
4) Encouraging annual cropping
5) Reducing fruit load and encouraging more vigorous growth of the tree
6) Aiding natural thinning that may have been insufficient

Fruit that is most often hand thinned are apples, pears, peaches and plums. Cherries do not seem to respond to or benefit from hand thinning. Since each fruit has its own thinning requirements, hand thinning is discussed in detail in the chapters on the individual fruit.

In summarizing the process from flower bud development to harvestable fruit, we see that the process really takes place over the course of two growing seasons. First, the flower or fruit bud is developing internally during mid summer of the first growing season. Following a rest period, during which cold temperatures are required, the bud continues to form slowly. Then, in spring, the flower blooms, male pollen is transferred to the female stigma, and the tree's genetic material is united to begin formation of the pit or seeds of the fruit. If all processes have occurred successfully, fruit begins to form on the tree while excess or damaged fruit is shed or thinned. Depending on the fruit, a harvestable crop will be ready anywhere from about 60 days to 150 days after blossom.

Section IV.

Pests & Diseases

14. Insect Pests

No doubt one of the most frequent questions asked by backyard fruit growers is "what is eating my tree's leaves or chewing on my fruit?" Because of their easy visibility, these effects of insect presence are usually first to be noticed. Chewing on fruit or leaves can certainly do serious damage to your crop and as a result the adversary relationship between gardener and insect often begins. One needs to realize, though, that not all insects are bad. Many are actually very beneficial; and some exist in the garden with little or no consequence.

In this chapter you will learn to identify the insects, both good and bad, most common to fruit trees. Chapter 16 will provide information on how to control or encourage these insects. Since many insects change form during different stages of development, they are sometimes difficult for the novice gardener to identify. Consequently, it is helpful to understand insect life cycles. This understanding may also help you to identify an unfamiliar insect.

Understanding Insect Life Cycles

All insects and mites go through similar and predictable stages as they mature from the egg to the adult. Basically this cycle can be diagrammed as follows (Figure 51):

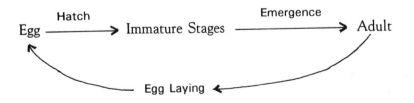

Figure 51. General insect life cycle

All insects start as eggs, laid by the adult female. She may lay a single egg or mass of eggs. Under favorable conditions, the egg hatches and goes through several immature stages. At the end of this growth phase the insect emerges as an adult. Mating takes place and the cycle is repeated. For some insects, it may take a full growing season, from spring to fall, to go through one complete cycle. Others may complete the cycle in a matter of days and produce several generations within a single growing season. At certain times during this cycle, the insect is more vulnerable to control than at other times. A look at these different growth periods shows how you might take advantage of them to control unwanted pests.

The insect eggs can take many forms. Some are flattened ovals, some are set up on stalks and some may be barrel shaped. They may be laid individually (such as aphid, mite, or fruit fly eggs) or they may be found in compact groups surrounded by protective coverings (such as tent caterpillars or leaf rollers). Eggs are laid in numerous locations such as the underside of a leaf, inside the developing fruit flesh, under bark scales, and in the soil.

Once the egg hatches, it goes through a number of growth stages called instars. Each instar is separated by a phase called molting. Since insects have no skeletons to support their bodies, they are supported and protected by a hard, rigid exoskeleton or outer skin. Unlike vertebrate skin, the exoskeleton does not expand as the insect develops. To grow, the insect must shed its exoskeleton and develop a new, larger one. This process is called molting. For a short time while the insect is shedding its old skin and before the new skin is fully developed, the insect's soft body is exposed and quite vulnerable to injury. This is often the stage at which it is easiest to control the unwanted insect.

When the insect molts, it not only grows larger, its body may change form also. This process of changing form is called metamorphosis. In some insects the adult form closely resembles the immature form, which is also known as the nymph or larvae. The only differences may be that the adult has wings and reproductive organs and is larger in size. This type of minor change in form is known as gradual metamorphosis. Insects that undergo gradual metamorphosis include plant bugs, aphids, mites, leafhoppers and scale.

Other insects undergo a process of complete metamorphosis. The physical change from the immature form to the adult is obvious and dramatic. In complete metamorphosis, the insect develops from an

Gradual Metamorphosis

Egg ———→ Several Larval Stages ———————→ Adult
                    (instars)

Complete Metamorphosis

Egg ———→ Larval Stages ———→ Pupa ———→ Adult
                 (instars)

Figure 52. Gradual and complete metamorphosis compared

egg through one or more larval stages. The larva then undergoes a resting stage known as the pupa. When it emerges from the pupal stage it is completely transformed as an adult. Beetles, flies and moths exhibit complete metamorphosis.

Insects that undergo gradual metamorphosis, as well as appearing similar in form, tend to spend their whole life cycle on the same host plant. Those that undergo complete metamorphosis however, may inhabit very different locations and exhibit different feeding habits with different life stages.

Insect Damage

Damage caused by insects can be of two kinds. Direct damage to the fruit is usually from chewing on the fruit, laying eggs in the fruit, burrowing within the fruit flesh or in some other way visibly disfiguring the fruit. Indirect damage, however, is more subtle and involves damage to the tree or its leaves. Insects will suck nutrients from the tree, bore into the tree, or reduce the tree's food producing leaf surface by injuring the leaves. The damage weakens the tree and makes it more difficult for the tree to properly photosynthesize. Indirectly, your harvest is diminished by smaller or fewer fruit.

General Insect Identification

Body color and shape are the most commonly used traits for identifying insects. It helps to be able to identify the basic body parts too. Both immature and adult insects bodies are made up of three basic areas; the head, the thorax and the abdomen. The illustration

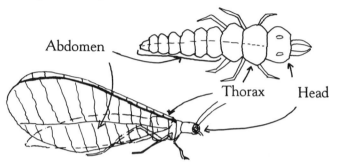

Figure 53. *Basic insect body parts*

shows various insects stages and their body part arrangements. Basically, the head has the sensing organs (eyes, antennae, mouth). The legs and wings are attached to the thorax and the abdomen carries the sex organs. In the immature stages, some body parts such as sex organs or legs may be absent and in some cases the abdomen may have leg-like appendages. Some immature forms may also have very small heads. To identify your insect, observe carefully. Ask your self these questions:

1) What color is its head and body?
2) How many legs does it have?
3) Does it appear to be adult or immature?
4) Are its wings and body hairy or hairless?
5) Does it have visible antennae?

By answering these questions and comparing your specimen insect to the illustrations in this chapter, you should be able to identify the group into which it belongs. In many cases, you will also be able to identify it more specifically by name.

The illustrations on the following pages show the most common insect pests that the backyard orchardist is likely to encounter. Certainly there are others which appear only occasionally or only in certain regions of the country. Your local cooperative extension service should be able to help in their identification.

Direct Insect Pests

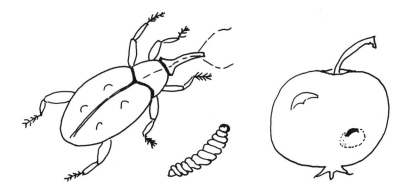

Name: Plum Curculio
Description: Grayish brown or black mottled, 1/5" long snout nosed beetle with 4 humps on back. Larvae are whitish yellow color.
Host Plant: Apple, plum, cherry, peach, apricot; occasionally pear.
Damage Done: Adult lays egg under skin leaving characteristic crescent shaped scar, larvae burrow into and feed on fruit flesh; damaged fruit usually drops from tree.

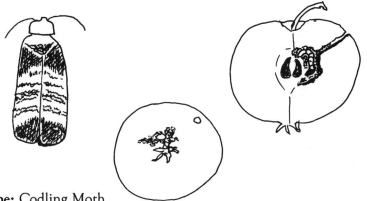

Name: Codling Moth
Description: Charcoal grey head, mottled grey wings with copper band on end. Adult moth ½" long. Larvae are one inch long with very pale pink to white body and brown head.
Host Plant: Apple, pear; occasionally cherry.
Damage Done: Larvae burrow into fruit and feed around core causing premature fruit drop. Frass often surrounds tell-tale exit hole at calyx.

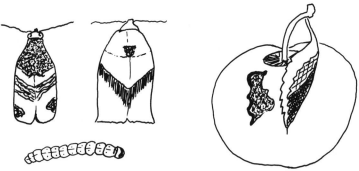

Name: Leafroller (redbanded, obliquebanded, fruittree, pandaemas)
Description: White with reddish band on wing, reddish brown or brownish tan mottling, ½" to 1" long. Larvae of various colors - white, yellowish, green, depending on species.
Host Plant: All fruit trees, primarily apple and pear.
Damage Done: Larvae chew on fruit surface, buds, and leaves. Spin webs around leaves to form tightly curled feeding shelter.

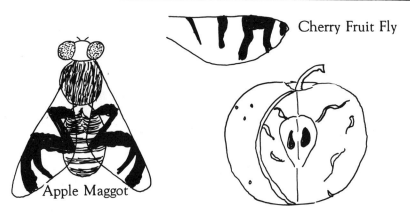

Cherry Fruit Fly

Apple Maggot

Name: Apple Maggot/Cherry Fruit Fly
Description: Black head, characteristic black wing pattern on adult. Larvae are legless, 1/3" long, yellowish-white in color.
Host Plant: Apple, cherry, plum
Damage Done: Eggs laid under skin, larva burrows into flesh and leaves "railroad" pattern of brown tunnels in flesh. (Small pinpoint sting visible on fruit surface.)

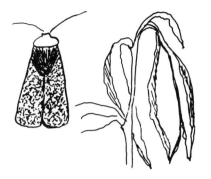
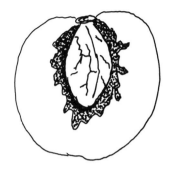

Name: Oriental Fruit Moth
Description: ½" long moth with grey mottled wings; larva white body with brown head, ½" long.
Host Plant: Peach; occasionally cherry, plum.
Damage Done: First generation larvae burrow into twig tips, killing twigs. Later generations burrow into fruit. (Early season peaches slightly less prone to Oriental Fruit Moth damage.)

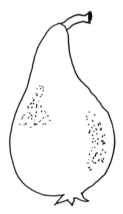

Name: Pear psylla
Description: Transparent yellow brown, 1/8" long "jumping louse". Nymph is wingless, adult has wings.
Host Plant: Pear, found mostly on underside of leaf and petiole.
Damage Done: Sap feeding weakens tree, "honeydew" from feeding becomes a growth media for sooty mold on fruit.

Insect Pests 167

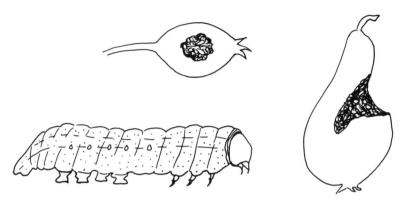

Name: Fruit Worms
Description: Steel or bluish gray moth. Larvae up to 1" long, various colors depending on species; usually green or brown with white speckles and body stripes.
Host Plant: All tree fruit, except peach.
Damage Done: Larvae feeding on leaves and fruit can badly disfigure fruit.

Indirect Pests

Name: Mites (European red, Rust, Two spotted, several others)
Description: Pinpoint size, color varies from red to translucent light yellow to cream. Usually found on the underside of leaf.
Host Plant: Apple, pear, cherry, plum, peach, apricot.
Damage Done: Sap feeding can weaken tree predisposing it to other problems. Rust mite causes brown russetting on apple and pear skin.

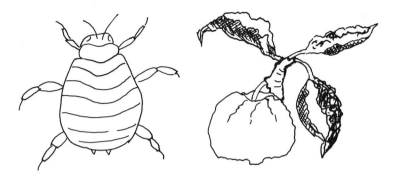

Name: Aphids
Description: Colors vary - red, green, black, gray, yellow. Less than ¼" long.
Host Plant: All tree fruit, depending on aphid species.
Damage Done: Suck sap form leaves causing leaves to curl; can severely weaken young tree. Population can multiply quickly. Sooty mold grows on sticky "honeydew" excreted by aphid.

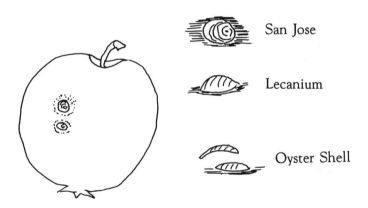

San Jose

Lecanium

Oyster Shell

Name: Scale (San Jose, Oyster shell, Lecanium)
Description: Light tan to grayish 1/16" long scale covers immobile developing immature stage. Mostly found on bark; visible on fruit too, when populations are high.
Host Plant: Most tree fruit, different species prefer different fruit.
Damage Done: Sap feeding weakens tree if populations are high, also scale is difficult to remove from fruit and is unsightly.

Name: Borers (Peach Tree, American Plum, Lesser Peach Tree)
Description: Adult is clear winged (except Am. Plum) moth with black wing pattern. ½" to ¾" long.
Host Plant: Peach, apple, plum, cherry in some areas.
Damage Done: Boring into trunk can weaken tree. On dwarfing apple rootstocks, will often bore in at graft union.

Occasional Insect Pests

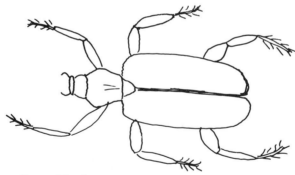

Name: Rose Chafer
Description: Very motile winged beetle; 1/2" long; tan wings with reddish brown edges; black underbody; long, thin, hairy legs.
Host Plant: Peach, plum, cherry.
Damage Done: Skeletonize leaves and flowers. Present in large numbers in June and July. Worst on sandy sites adjacent to grassy areas.

170 the Backyard Orchardist

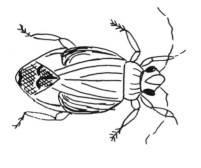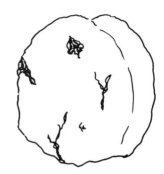

Name: Tarnished Plant Bug
Description: Yellowish-brown mottled wings; 1/4" long, may have black spots or reddish stripes on body.
Host Plant: Peach (also called catfacing insect), apple, pear, cherry, plum.
Damage Done: Inject toxins into buds and shoots, causing dwarfed shoots and sunken areas (catfacing) on fruit.

Many insects will produce more than one generation of offspring over the course of each fruit growing season. How many generations are produced is somewhat dependant on the length of the growing season in a geographic area. Table 26 indicates what you can typically expect to observe. Also, geographic location can dictate which insect species are commonly found there. The map in Figure 67 gives an indication of those pests most likely to be found in each region.

Table 26 - Insect Generations per Growing Season

Insect Pest	Generation	Insect Pest	Generation
Plum Curculio	1 or more	Codling Moth	2 or more
Leafrollers	2	Cherry Fruit Fly	1
Apple Maggot	1	Oriental Fruit Moth	3 to 4
Pear Psylla	1	Fruit Worms	1
Mites	6 to 8	Aphids	3 to 4
Scale	2	Borers	1
Rose Chafer	1	Tarnished Plant Bug	2 to 5

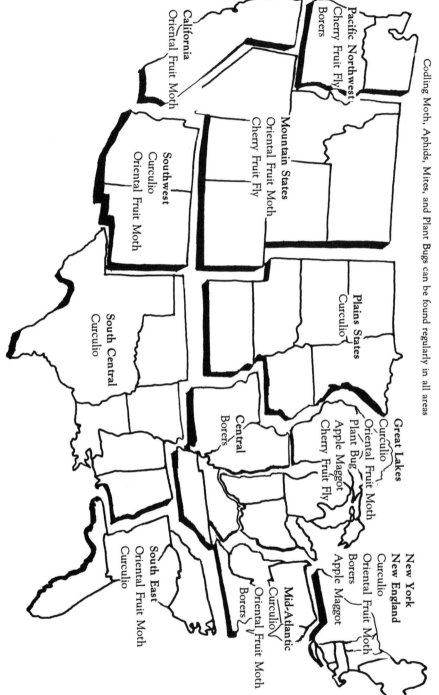

Figure 67. *Common tree fruit insect pests in various regions of the United States*

Codling Moth, Aphids, Mites, and Plant Bugs can be found regularly in all areas

Pacific Northwest
Cherry Fruit Fly
Borers

California
Oriental Fruit Moth

Mountain States
Oriental Fruit Moth
Cherry Fruit Fly

Southwest
Curculio
Oriental Fruit Moth

South Central
Curculio

Plains States
Curculio

Central
Borers

Great Lakes
Curculio
Oriental Fruit Moth
Plant Bug
Apple Maggot
Cherry Fruit Fly

New York
New England
Curculio
Oriental Fruit Moth
Borers
Apple Maggot

Mid-Atlantic
Curculio
Oriental Fruit Moth
Borers

South East
Oriental Fruit Moth
Curculio

172 the Backyard Orchardist

Beneficial Insects

 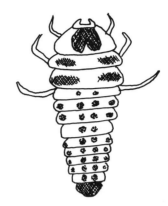

Name: Ladybug Beetle
Description: Adult shiny black, red, or orange. 1/16" long. Larva blue/black, reddish grey, 1/8" long. Color varies with species.
Insects preyed on: Aphids, scale, mites

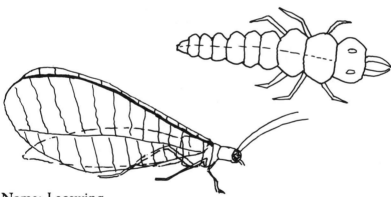

Name: Lacewing
Description: Adult translucent whitish green with copper eyes. Larva light tan with long moth parts.
Insect preyed on: Aphids, leafhoppers, mites, scale, and certain moth eggs including Oriental Fruit Moth

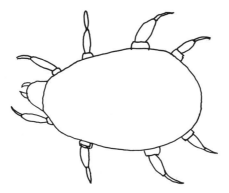

Name: Predator Mites (several species)
Description: Translucent tan to light red.
Insect preyed on: Plant feeding mites

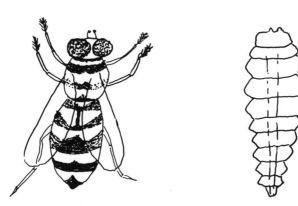

Name: Syrphid Fly
Description: "Hovering" fly looks much like a bee or yellow jacket wasp. Larva pale green to tan, "slug-like" appearance.
Insect preyed on: Aphids, scale

174 the Backyard Orchardist

15. Disease Identification

Fruit tree diseases fall into three major classifications based on what causes them. There are diseases caused by fungi, bacteria, and viruses. In all three cases, the actual organism causing the disease is hard to view with the naked eye, but the disease symptoms become rapidly apparent if allowed to progress. For most diseases, prevention is a more successful means of control than after the fact eradication. Therefore it is helpful to understand the conditions that bring on the various diseases and the conditions under which they flourish.

Fungus Diseases

Let us look first at those diseases caused by fungi. All the fungus organisms share a characteristic cycle of growth and reproduction. The general cycle is illustrated in Figure 72.

Most of the major fungus diseases common to tree fruit have both a primary and a secondary infection period during which they may cause problems for your fruit or fruit tree. Overwintering leaves or fruit that have dropped to the ground often serve as host sites for fungal reproductive structures. During the primary infection phase, primary spores, or ascospores, are released from the reproductive structure. Under proper conditions the spores germinate and infect young developing flowers, fruit, or leaves. Secondary fungal fruiting structures are soon produced. If effective disease control measures are not applied during the primary infection phase, secondary fungal spores are released from the fruiting bodies and additional secondary disease infection of the fruit and leaves is likely. Following secondary infection, diseased leaves and fruit often drop to the ground and overwinter. The following growing season the cycle is repeated again.

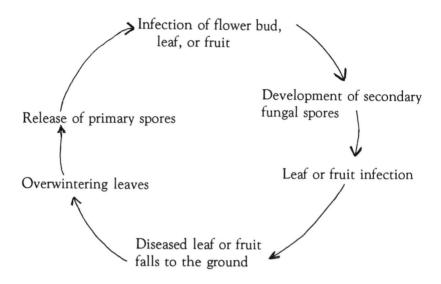

Infection of flower bud,
leaf, or fruit

Development of secondary
fungal spores

Release of primary spores

Leaf or fruit infection

Overwintering leaves

Diseased leaf or fruit
falls to the ground

Figure 72. *Generalized fungus life cycle*

In order to grow and continue reproducing, all fungi need certain favorable conditions. These include moisture, temperature, and a host (in this case, your fruit tree). In general, fungus diseases are most active in the temperature range present during the spring and summer growing seasons. When adequate moisture (from rain or high humidity) is present, the fungus will grow and infect either the tree or the developing fruit (and in some cases both). Different fungus diseases can cause mold and rot of the fruit, reduce the tree's leaf area and photosynthetic ability, or cause distorted growth of the branches, leaves or fruit. Some of the most common fungus diseases of fruit are found on the following pages.

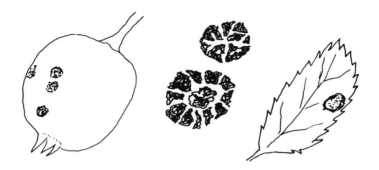

Name: Scab
Description: Brown irregular spots on leaves; dark brown or black corky growth on fruit. Severely infected fruit may be misshapen.
Host Plant: Apple, pear
Damage Done: Reduces leaf photosynthesis. Disfigures fruit.

 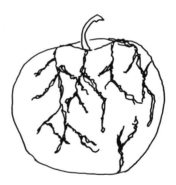

Name: Powdery Mildew
Description: Leaves covered with powdery white fungus netting, may curl and crinkle. Tan netting on severely affected apples.
Host Plant: Apple, cherry
Damage Done: Impaired photosynthesis.

Disease Identification 177

Name: Brown Rot
Description: Fruit turns brownish with round white/brown fungus spots on surface.
Host Plant: Peach, cherry, plum, apricot
Damage Done: Fruit soft, rotten and unusable. Disease can multiply rapidly.

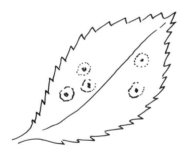 

Name: Cedar Apple Rust
Description: Yellow orange spots on underside of leaf and calyx end of fruit. May have tube-like growths growing from spots. Yellow orange horned structures seen on alternate hosts.
Host Plant: Apple, (hawthorn, Eastern red cedar, juniper are alternate hosts)
Damage Done: Raised blister on fruit and leaves. Can defoliate severely infected tree. Complete life cycle spans 2 growing seasons.

178 the Backyard Orchardist

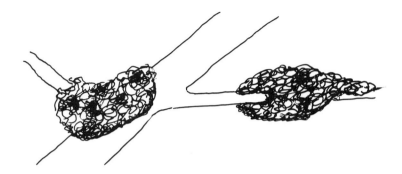

Name: Black Knot
Description: Hard, black, knobby looking fungus overgrowth on twigs and branches.
Host Plant: Plum, occasionally cherry
Damage Done: Can girdle and kill branches.

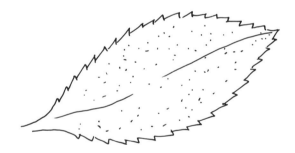

Name: Leaf Spot
Description: Small purple/brown spots on leaf. Leaves turn yellow and drop from tree.
Host Plant: Cherry
Damage Done: Lesions on leaves impair effective photosynthesis, trees lose leaves.

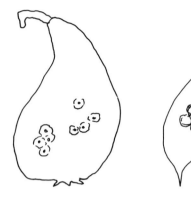

Name: Fabraea Leaf Spot
Description: Small, round, purplish lesion with black in middle.
Host Plant: Pear
Damage Done: Impaired photosynthesis, trees lose leaves, fruit misshapen and cracked.

Name: Bitter Rot and White Rot ("Bot" rot)
Description: Small brown sunken lesions, may be surrounded by red halo. Rot works its way to core. In Bitter rot affected area is still firm, white rot area is soft.
Host Plant: Apple
Damage Done: Fruit rots to core.

180 the Backyard Orchardist

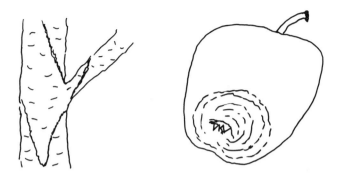

Name: Black Mold (Frog Eye Leaf Spot)
Description: Concentric rings of brown/black rot on calyx end of fruit. Reddish brown, dry, sunken cankers on limbs
Host Plant: Apple, pear
Damage Done: Fruit rots and is unusable.

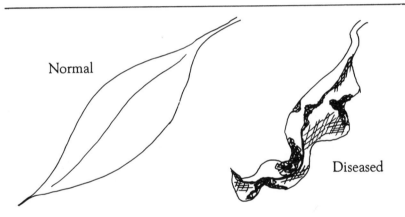

Normal

Diseased

Name: Leaf Curl
Description: Leaves become thickened and pucker up. Reddish purple coloration along puckered area. Leaves turn yellow and drop.
Host Plant: Peach
Damage Done: Trees defoliate when severely infected.

Bacterial Diseases

Some diseases of fruit are caused by bacteria. Under favorable conditions, bacteria can multiply very quickly. As a result, the diseases caused by them can be very damaging if allowed to develop uncontrolled.

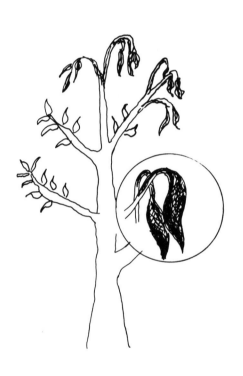

Figure 83. Fireblight

Fireblight is one of the most serious diseases of susceptible apple and pear varieties. As its name implies, the infected trees sometimes look like they have been burned by wild fire. Major shoots and even whole trees' foliage can turn black and wilted in less than a week with severe infection pressure.

A general rule of thumb - "the 65/65 rule", says that fireblight will develop if temperatures are above 65° F and relative humidity exceeds 65%. Open blossoms and tender, rapidly growing vegetative shoots are particularly susceptible. Visible disease symptoms show up on the tree within three to four days of infection. Home gardeners should also be aware that fireblight is not just a disease of fruit trees. It can also attack European Mountain Ash, Hawthorn and several other landscape plants.

Control methods for fireblight include planting varieties that are resistant (none are truly immune), spraying the trees with streptomycin sprays during periods of infection, controlling sucking insects that spread the disease, and pruning out infected wood. When pruning blighted limbs, it is best to do so during the dormant season when fireblight bacteria are not active. Cut nine to twelve inches below the visible disease canker into healthy wood. This is especially important if the tree is growing actively. Be particularly careful not to spread ooze from the canker around as it contains large amounts

of active bacteria. Re-
move infected prunings
from the garden and burn
them. Past recommenda-
tion has been to carefully
disinfect pruning tools in
a 3% hydrogen peroxide
or 10% bleach solution.
New research is indicat-
ing that this practice may
be of limited value and
considering that bleach is
very corrosive, you may
find the dormant pruning
is the best solution.

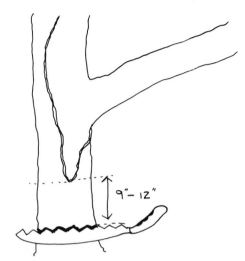

Bacterial Spot can be
a severe problem on
certain varieties of apricot, nectarine, peach,

*Figure 84. Bacterial ooze around fireblight
canker and proper location of pruning cut*

and plum. A common problem throughout the eastern United States,
the disease first shows up on the undersides of leaves as a number of
water soaked black or brown spots. Frequently the center of the
infected spots fall out and a hole with a red halo remains. Badly
infected leaves turn yellow and fall from the tree. This can cause
weakening of the tree and result in smaller fruit.

Infected fruit takes on an unsightly appearance with cracked
skin and sunken areas in the flesh.
Infection is most severe during
rainy periods of moderate temperature and high wind. Most nectarines and apricots are susceptible to
bacterial spot.

Chemical controls have not
been completely effective in halting
the disease's spread. Maintaining
healthy trees that have not been
given excess nitrogen and choosing
resistant varieties are the gardener's
surest means of control.

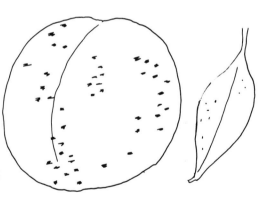

*Figure 85. Bacterial Spot on leaves
and fruit*

Disease Identification   183

Bacterial Canker is most common on sweet cherry trees, but can also infect peach, plum, and tart cherry. Symptoms of the disease include cankering and heavy gumming of infected limbs. On sweet cherry, infected leaves roll up and show spotting along the edges. Sour cherry leaves may yellow and show dead spots where tissue has dropped out of the leaf. Infected fruit develop dark, deeply sunken areas. These fruit are also more susceptible to the brown rot fungus.

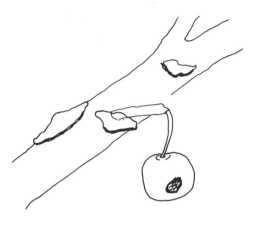

Figure 86. Bacterial canker

Periods of cool wet weather after bloom favor development of bacterial canker. Sprays of copper in spring and during the early growing season help control the disease.

Crown Gall bacteria infect the roots of many fruit as well as other landscape trees. The disease develops as small, tan colored rounded tumors on the root system. Eventually crown gall tumors become several inches in size, hard, and woody. Most often they end up girdling the crown or large main roots.

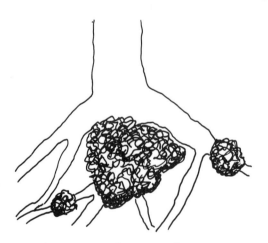

Figure 87. Crown gall on roots

Crown gall is found most often in soil where susceptible species have been intensively grown in the past. Tree nurseries can become a ready source of infection. Peach root and mazzard cherry show a high incidence of crown gall although they are not the only fruit trees that can become infected. The best control method is to purchase disease free nursery plants. Roots can also be dipped in a solution of a noninfecting, competitive bacterium called *Agrobacterium radiobacter* if planting on a larger scale.

Virus and Miscellaneous Disorders

Virus diseases tend to develop more slowly than bacterial diseases, sometimes taking several years before noticeable symptoms occur. Often, virus disease symptoms may mimic other problems such as poor tree nutrition, fungus diseases, or general weakness of the tree, which makes viruses harder to identify. Much research is still being done on virus diseases and hopefully, in time, identification and control will become easier. The few virus or virus-like conditions that the home gardener is prone to encounter are listed below.

Stony Pit Virus of pears is found most often in Bosc pears. Anjou and Winter Nelis are also affected by this disease that can cause severe deformity of the fruit. Symptoms appear quite similar to stink bug injury and are often mistaken for it. Heavily infected fruit will be badly gnarled. A number of hard little "stones" will be found in the flesh. When cut with a knife, the flesh will feel gritty.

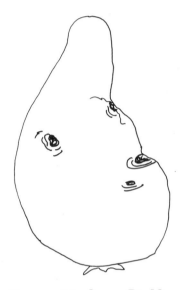

No cure is known for Stony Pit Virus. the best prevention is selecting resistant varieties or purchasing known virus free trees.

X-Disease of peach, nectarine, and cherry is found mostly in areas with wild chokecherry populations.

Figure 88. Stony Pit Virus

Caused by a virus like organism called a mycoplasma, X-disease develops over several years and is easiest to identify on peach trees. Isolated limbs will have leaves with yellow to purplish spots. The leaves soon curl up and drop from the tree. Soon only a cluster of leaves is left on the end of the infected branches. After several seasons, symptoms show up on most limbs.

On cherry, the symptoms may show up as yellowish unripe fruit mixed in with normal fully ripe fruit. Leafhoppers sucking sap from trees appear to be a major means of spreading X-disease. Controlling the leafhoppers is second only to removal of neighborhood wild chokecherry as a means of controlling the disease.

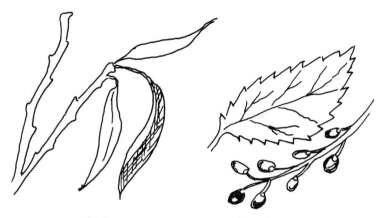

Figure 89. X-Disease symptoms and chokecherry alternate host

Nematodes

Nematodes are not a fruit tree disease, nor are they an insect. They are a microscopic round worm that lives in the soil. Since nematode problems often mimic fruit tree disease symptoms, they will be discussed in this chapter.

Nematodes are rather primitive worms that feed as parasites on plant roots. Sometimes their feeding will weaken the roots sufficiently to allow secondary diseases to enter the tissue. If your fruit tree is growing poorly and you have ruled out other disease and nutrition causes, consider testing the soil for nematodes. Samples are taken much like the soil nutrient samples and can often be done at the same time. Your county extension office can advise on the proper procedure for your area.

If you are able to examine the roots of your tree, you may see nematode feeding symptoms. They include root galls, dead areas of root tissue, injured root tips, excessive branching of roots or generally stunted growth of the roots or tree.

Several methods of nematode control exist. Choosing resistant rootstocks when possible is the most feasible for the backyard fruit grower. Soil fumigation or starving the nematode population with a cover crop for two or three seasons is used in commercial orchards, but is not very practical in the small space of the home garden.

16. Insect & Disease Control Methods

In the past most gardeners had the impression that insect and disease control for fruit trees could only be accomplished by spraying some sort of chemical. With increasing awareness of organic gardening and new developments in integrated pest management several options are now available. Which method you should use is a choice you will have to make. Chemical control methods are, at first glance, the easiest and quickest for the novice gardener. Organic gardening and integrated pest management take more time, knowledge, and observation, but are becoming increasingly preferred by many for their environmental benefits.

Since there are many new names and terms associated with present pest control philosophies, let's first define the ones to be used in this chapter.

Chemical control has been considered by many as the mainstream pest control method in recent decades. This method has assumed that all "bugs" and diseases are problems to be eliminated. In this book, chemical materials are considered to be those manufactured or extensively processed as opposed to those left in their fairly natural form. Advantages of chemical use include quick, effective control of insects and diseases with a minimal expenditure of labor. It is also responsible for the attractive abundance of unblemished produce found on our grocery shelves. Undesirable consequences of this method include insects and diseases developing resistance to certain chemicals, as well as ecological ramifications, such as ground water pollution. Accidental poisoning of the person applying chemicals and possible long term health problems are also risks of chemical control systems.

With present day ecological awareness, organic pest control is much talked about. Just what exactly is organic gardening? Justifiably there is some confusion as individuals and various groups each have their own definitions. As someone once defined their understanding to the author, "unsprayed and unfertilized", is a simplistic definition. On the other end of the spectrum are organic growers groups that do allow selective use of certain "chemical" pesticides in their certification programs. For purposes of this book, organic strategies will be defined as those that use primarily naturally occurring materials in relatively unprocessed form. This is not to say that all these materials are good or safe and that man-made ones are all bad or dangerous. Excessive manure applications can cause just as much water pollution as manufactured fertilizers. Some organic pesticides, such as nicotine sulfate, are more poisonous than manufactured ones, such as Imidan.

Growing fruit organically requires that you have a thorough understanding of the growth processes of the tree, the fruit, the pests, and the beneficial insects; and how they are all part of an interrelated cycle. The major advantage of this system is that it works in harmony with the natural system. Disadvantages are that it can be labor intensive and fruit may not be as picture perfect and blemish free as many people have become used to. Research so far has shown that tree fruit are one of the most difficult crops to grow under a strict organic definition. Several pests, most notably curculio, are not effectively controlled by any currently available organic means. Consequently, a large percentage of crop loss to insect damage is something you may have to accept in growing tree fruit organically.

Integrated Pest Management (IPM for short) is a strategy that is rapidly gaining acceptance. IPM also requires the grower to have a thorough understanding of the plant, the pests, the beneficials and the ecosystem. In many ways, it is a middle ground between chemical control and pure organic control methods. Practiced properly, IPM will allow you to quantify and keep a history of your plant, its problems and its performance. This sounds complicated and like a lot of work, but it actually allows you to eliminate unnecessary work. It will also lead you to a better understanding of the natural balance. Integrated pest management takes the advantages of both chemical and organic methods, weighs them against the disadvantages, and applies them in the context of an interdependent system.

An integrated pest management system allows you to decide what trade-offs you are willing to make. To limit the amount or type of pesticides you use, you may be willing to accept more blemished fruit or a few worms. Perhaps you would rather use a less toxic "chemical" pesticide than a highly toxic "natural" one. The choices are yours when you have sufficient information. Most of the pest control strategies discussed in this book have an IPM focus. If your personal philosophy leans heavily toward either organic or chemical methods, by all means follow your chosen path. Consider though, that IPM will still have much to offer you that is of value.

Safe Handling of Pesticides

All too often, safety in handling and application of pesticides is something that is glanced over or left out entirely in gardening books. It may not be something glamorous to think about, but do remember that a poison by any other name is still a poison. Whether it is a manmade chemical or an organic material it requires respect and a dose of responsible common sense. Most unintentional pesticide poisonings happen when material is used without proper knowledge or protection. In many states it is now necessary to be licensed to buy and apply the more toxic pesticides. If you choose to use these materials, pass the licensing exam and even if you are using the less toxic materials handle them with care and respect. As minimum common sense safety precautions:

1) Read the label carefully and know what you are using. Use only the recommended rate of material. **Follow the label directions carefully.**

2) When ever possible, mix up only as much spray as you will need for your current application. If you have a small amount of material left after spraying, it is better to give your fruit tree a bit heavier spray application. Through exposure to the weather the spray materials are most easily broken down naturally. **Do not dump leftover spray in a concentrated spot on the ground or down a sewer.** It will be more difficult for it to properly break down there.

3) Store pesticides out of reach of children, in a well marked, locked cabinet.

4) **Never** save left over spray materials in something other than its original container.

5) Spray only under calm conditions when spray will not drift onto neighboring property, pets, or yourself.

6) Dress properly. Wear long sleeves, pants and work shoes; **No** bare feet, sandals, shorts or bathing suits.

7) Use protective clothing (rubber gloves, boots, respirator, face

shield) when called for.

8) Do not eat, drink, or smoke while using pesticides. Wash carefully when finished.

9) Dispose of containers properly according to accepted local recommendations. This usually will mean rinsing the containers three times with clean water and disposing of them at a properly licensed landfill.

10) Follow the recommended interval from last spray to harvest.

Effective Pest Control

One of the main concepts behind both the integrated pest management approach and the organic gardening approach is that not every occurrence of a fruit tree pest or disease demands instant control. Some pests can be tolerated and some diseases prevented. If however, pest or disease conditions become excessive, some type of control method often becomes necessary. All effective control programs share certain basic principals. They include:

1) Planning for problem prevention
2) Proper identification of the pest or disease
3) Selection of an effective control agent - whether that be a predator insect, organic pesticide, or chemical material
4) Proper timing in application of the control method

If you have come to a point where you feel control is needed, your next step will be to properly identify the pest or disease. You should be able to identify most common problems after reading the two preceding chapters. Your local extension agent can help identify more unusual problems.

Once you have identified your pest, you will need to choose your control method. In most cases, this will be some type of pesticide, but it could also be some type of predator insect. Pesticide, (-cide meaning "to kill"), is the term generally used to define a material that kills pests - be it an insect, disease, or weed. The materials can generally be broken down into the categories of insecticide, fungicide (to kill fungi responsible for many diseases), bactericide or herbicide (to kill weeds or "herbs").

Choosing a material targeted specifically for the purpose of controlling your given problem is preferred in integrated pest management. It is often the most effective and, if chosen properly, is not injurious to other desired insects that also inhabit your garden.

Careful use of a selective pesticide can even be a valuable tool in letting beneficial insects gain control of a runaway pest population and bring it back into an ecologically balanced control situation.

The final necessity of an effective pest or disease control program is applying the selected material at a time when it will be most effective. The importance of proper timing can not be understated! Even the most effective material will not work well if it is applied at the wrong time. From Chapter 14 you are already aware that most insects are likely to be vulnerable to controls when they are immature or unprotected, such as when they are molting. Diseases are usually easiest to control before the bacteria or fungal spores have increased to great numbers.

Since each growing season is different, it is impossible to say that a given spray applied on a given date will always work. General recommendations, correlated to the tree growth phases, as diagrammed on pages 192 and 193, can be more reliable. As you keep records and gain experience, you will be able to fine tune your timing. The charts show the critical control periods for some of the more common insect pests and diseases in the northern United States. In the south, the sequence will still be the same, but may occur several weeks earlier. Use them as a guide when observing your fruit tree and apply control measures only when actually necessary.

No doubt some of you are wondering about the "all purpose" fruit tree spray you have seen or maybe even used. Isn't that the easiest control solution? It does seem easy; but it may not always be effective. The all purpose spray is usually a mixture of several broad spectrum insecticides and fungicides. If applied on a regular weekly basis, it will certainly kill a number of insects and stop some fungus growth. Unfortunately, it will also kill a number of helpful insects that might otherwise have kept pests under control for you without a spray. In most cases it will actually cause an enormous increase in damaging mite populations because it is very toxic to their natural predators. It also will not control serious bacterial diseases such as fireblight, since it contains no materials aimed at bacteria control. If you still feel you prefer to use the all purpose spray, realize that you will need to spray regularly (most likely weekly). You will probably be adding unnecessary chemicals to your food and the environment, and you will not always get all the disease or insect control that you would with a properly timed selective material.

Periods of Active Insect Pressure

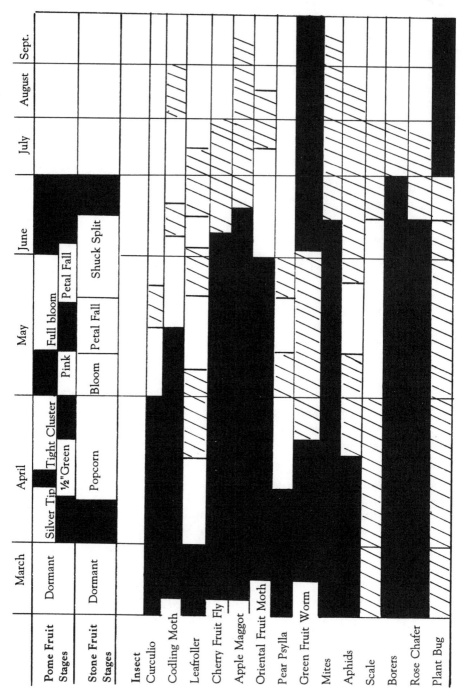

Figure 90. *Critical control timing for common fruit tree pests*

Periods of Active Disease Pressure

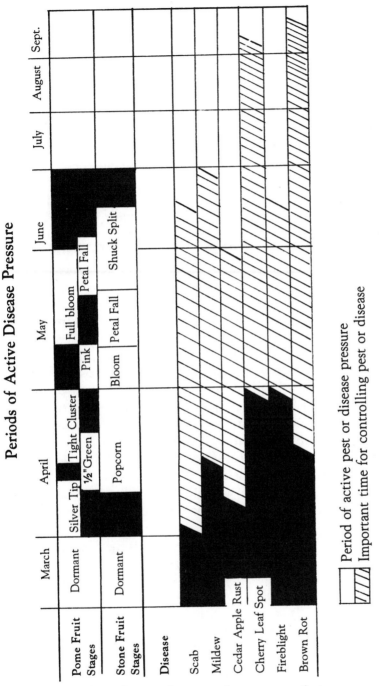

	March	April		May		June	July	August	Sept.
Pome Fruit Stages	Dormant	Silver Tip / ½"Green / Tight Cluster	Pink	Full bloom	Petal Fall				
Stone Fruit Stages	Dormant	Popcorn	Bloom	Petal Fall	Shuck Split				
Disease									
Scab									
Mildew									
Cedar Apple Rust									
Cherry Leaf Spot									
Fireblight									
Brown Rot									

☐ Period of active pest or disease pressure
▨ Important time for controlling pest or disease

Figure 91. Critical control timing for common fruit tree pests

Insect & Disease Control Methods   193

Insect Control

Chapter 14 helped you identify several of the major types of beneficial insects that can sometimes act to control pests. There are many species of beneficial insects and most are specific to given locations. Beneficials work by acting as parasites or predators of your pest insect. In order to flourish, they need enough pests or other food to feed on, a host plant to inhabit, and favorable weather conditions. Predator insects will normally thrive under proper conditions and, contrary to earlier beliefs, introducing predator insects into a home garden situation may not provide major benefits. Introducing a nonnative species can even cause hybridization and the development of a sterile population that is unable to produce the successive generations needed for continued pest control. In order for pest control by beneficial insects to be effective, the area protected must also be large enough that the pest species are not able to easily move into it. This often means several acres in size. When pest problems are not controlled by these natural predators, action is often needed to insure a harvestable crop. In most instances this involves spraying a natural or chemical control material on the trees and fruit.

Although the number of potential insect problems is large, it is unlikely that you will be bothered by more than a small number during any one season. Often, properly controlling the first generation of an insect infestation, reduces the amount of control needed (if any) later in the growing season. As mentioned earlier, one of the keys to effective pest control is selecting a material that will work to control your particular problem.

The chart that follows will give you a list of the currently available choices. Some materials may control more than one pest at a time and some may be more effective than others for a given pest. A notation in parenthesis indicates the effectiveness of the material against the listed pest as excellent (E), good (G), fair (F), or poor (P). Where no notation is made, you can assume the material does an average job of control, if it is applied at the proper time.

Some materials may come in several formulations, liquid, powder, granule, or emulsifiable concentrate. For proper application rates, consult the package since these may vary with the formulation of the material. Also check local laws. As more states and municipalities try to regulate chemical use, some material's use may be restricted or not allowed in certain areas.

Table 27 - Insect Control Options

Pest	Chemical	Organic
Curculio	Imidan (E) Malathion (P) Methoxychlor (P) Carbaryl (F)	none
Codling Moth	Imidan (E) Diazinon ((F) Malathion (P) Methoxychlor (F) Carbaryl (E)	Ryania (F) Rotenone (P) *Bacillus thuringensis*
Leafroller	Imidan (E) Carbaryl (F) Diazinon (P) Malathion (P)	*Bacillus thuringensis*
Apple Maggot Cherry Fruit Fly	Imidan (E) Diazinon (G) Carbaryl (G) Malathion (F)	Rotenone (P)
Oriental Fruit Moth	Imidan (E) Diazinon (G) Carbaryl (E)	Ryania (F)
Pear psylla	Diazinon (P) Malathion (P) Methoxychlor (P)	Dormant oil Diatomaceous earth Rotenone (P) Insecticidal soap (P)
Fruit Worms	Diazinon (F) Imidan (P)	

Mites	Kelthane (F)	Dormant oil Insecticidal soap (F) Diatomaceous earth
Aphids	Malathion (G) Diazinon (F)	Superior oil Insecticidal soap (F) Nicotine sulfate (F)
Scale	Malathion Diazinon (G)	Superior oil Insecticidal soap
Borers	Lorsban (G)	Hand removal by digging out of trunk with a knife Nicotine sulfate (F)
Rose Chafer	Carbaryl (G)	
Tarnished Plant Bug	Carbaryl	

Disease Control

The majority of spraying on fruit trees is done to prevent or eradicate diseases, which are often a season long problem. Disease pressure can in many cases be minimized by growing disease resistant varieties. Selecting these varieties is the biggest step you as a grower can take in reducing the amount of effort that you will have to invest in control later. Disease resistant varieties from some of the first breeding programs sometimes traded off fruit quality for disease resistance, but many of the latest disease resistant varieties now offer very good fruit quality. For lists of the best currently available disease resistant varieties, see the Chapters 4 through 9.

Unfortunately, even the best of the resistant varieties are not immune to all problems, so some form of disease control may be necessary from time to time. Prevention or early control of diseases is very important. The fungi and bacteria that cause most tree fruit diseases multiply rapidly in favorable conditions. Often, once the disease has infected the fruit, there is no way to eradicate it. In most cases, the only solution is to see that it does not progress further. The following chart outlines the major disease control options.

Table 28 - Disease Control Options

Disease	Chemical	Organic
Scab	Captan Benomyl Dodine Ferbam	Bordeaux (F) Sulfur (P)
Mildew	Benomyl (G)	Sulfur (F)
Brown rot	Captan (F) Benomyl Dodine Ferbam (F)	Sulfur
Cedar Apple Rust	Ferbam (F) Polyram (E)	Remove junipers & red cedar that serve as alternate hosts
Black Knot	Ferbam	Prune out and burn diseased limbs
Leaf Spot	Dodine	Copper sulfate at petal fall
Peach Leaf Curl	Ferbam (G)	Copper or lime sulphur in late fall after leaf fall or early spring before bud break
Fireblight	Streptomycin	Careful removal of infected wood Bordeaux or fixed copper and oil early in the season

Insect & Disease Control Methods 197

Bacterial Canker	Bordeaux or Fixed Copper plus oil early in growing season
Crown Gall	Dip tree roots in *Agrobacterium radiobacter* before planting

Notes regarding pest and disease control materials:

Do not use carbaryl within 30 days of bloom on apples, otherwise fruit thinning may occur.

Carbaryl and diazinon are highly toxic to bees. Do not use during bloom or other times when bees are active around your fruit trees.

Benomyl and Dodine should always be used in combination with another scab or brown rot fungicide. If used alone, development of resistant fungus strains is likely.

Alternate chemical choices throughout the season when ever possible to avoid development of pesticide resistant insects and diseases.

Do not use oil with or following Captan.

Sulfur will cause a rough fruit finish on apples, particularly on sensitive varieties.

Do not use oil within two days before or after temperatures of less than 40° F.

Do not use sulfur within 30 days of oil.

At temperatures above 85° F do not use either oil or sulfur.

Trade names for some of the materials listed include

Benomyl = Benlate or Topsin M

Carbaryl = Sevin

Carbamate = Ferbam

Dodine = Syllit

Bacillus thuringensis = Dipel B.t. (There are several strains of Bacillus thuringensis. Each is specific to different pests. Be sure to use the proper strain for fruit pests.)

17. Wildlife Pests

Along with controlling insects and diseases the fruit grower occasionally has to contend with other garden visitors as well. Fortunately, these pests are easy to identify and pose more of a nuisance in the backyard than a serious problem.

Birds

Birds in the home orchard can be a mixed blessing. Often they make a substantial contribution to controlling insects and are valued as a nonchemical pest control. At the same time, birds enjoy feeding on the juice of certain fruits. Cherries are particularly favored. Birds will also peck certain pear varieties, particularly those with a red blush.

Several deterrents are available. Most work for short periods. Experience has shown, though, that birds do learn which ones pose a real threat and soon ignore any one deterrent, if used for an extended period. The best strategy is to rotate methods.

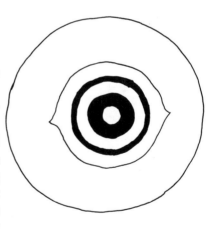

One of the most effective bird deterrent methods is to cover the fruit tree completely with netting several weeks before harvest so that birds cannot reach it. This works well but can be cumbersome if you are dealing with more than a few small trees.

Figure 92. Bird "Scare Eye"

Believe it or not, the old "scare crow" method of hanging a dead crow high on a pole in the orchard seems to have some effect for a short time. You can also try aluminum pie plates or strips of foil hung from string to wave in the breeze.

Wildlife Pests 199

The most recent technique in bird scaring comes from Japan and appears to be working quite effectively even in large orchards. Here a large inflated beach ball sized balloon is suspended from a pole about ten feet high. The balloon has been imprinted with a colored pattern that appears to the birds as the eye of a hawk. The balloons are readily available from many mail order orchard supply houses and work well to keep the birds from unwanted locations.

Deer

Deer are more of a problem for the gentleman farmer with country acreage than for the suburban gardener. They can, however, cause considerable damage to a young fruit tree, and do so in a hurry. In the early summer, deer enjoy feeding on the new shoot growth and in one or two bites can prune back a potential limb to nothing. Deer saliva also seems to be an excellent medium for transferring fireblight bacteria from one tree to another.

Various deer deterrents have proven themselves useful. Like bird deterrents, though, they need to be rotated, lest the deer become too familiar with one and ignore it. One very popular and easy option is to drill a hole through a hotel size bar of scented soap and attach it to the tree with a "twist tie".

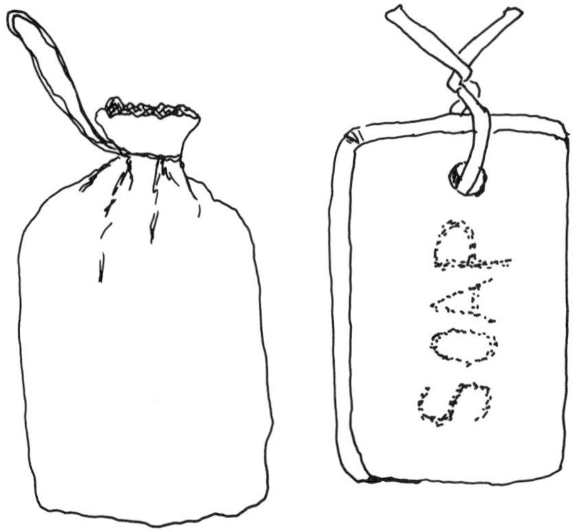

Little cloth bags with human hair or animal blood and bone meal, known as animal tankage, can be tied to trees. The scent keeps the deer wary and will deter them in areas where they have not become too domesticated from being fed by animal lovers. Tankage or hair bags will need to be changed every few weeks as the scent dissipates. You will also need to be vigilant that a branch does not become strangled

Figure 93. Hotel soap bars & tankage bags ready to hang as deer deterrent

where the bag has been tied. (If possible, hang the bag on a branch you know you will later prune out, or purposely leave a stub for it.)

A commercially prepared spray of ammonia soap (Hinder™) or putrefied eggs (Deer Away™), is another possibility. This is not as foul as it sounds. Although the odor is quite strong to the deer, we humans have an inferior sense of smell in comparison and rarely detect any odor unless we purposely stick our noses in the tree. Even then the odor is mild. A word of caution on this type of product though. It can burn tender young foliage and should be applied only to dormant trees. To use during the growing season, soak a sponge in the material and attach it to a stake near the tree. If the stake is about deer nose height it does an effective job. The sponge will need to be resaturated periodically, especially after heavy rains.

Fencing, although used effectively in commercial orchards, is a final option. Normally it will be too costly for the small garden. If you must resort to this, remember that a deer is perfectly capable of jumping a nine foot fence or slithering under a wire as low as only two feet high.

Rodents

Mice, voles, and rabbits can play real havoc with fruit trees. Most of their damage is done in the winter. Under cover of snow, they will gnaw on the trunk, often chewing away a wide ring of bark all the way around the trunk. The girdled trunk can be difficult to repair when discovered several months later. The best cure for rodent damage is prevention.

Coiled plastic mouse guards serve the purpose well. They are easy to get through nursery suppliers, inexpensive and take little effort on the part of the fruit grower. Normally the mouse guard is installed just after tree planting. With a gloved hand (to avoid a sharp "paper cut"), stretch the guard upward to uncoil it. Place the base of the guard at the base of the tree and slowly release it so that it coils around the tree as it rewinds. Be sure the installed guard rests on the soil, but does not wrap around the trunk below the soil line. A guard that is trapped between the trunk and settled soil cannot expand as the tree trunk does and will eventually constrict the tree. Likewise a guard that does not extend down to the soil surface will allow mice to crawl up inside, defeating its purpose.

The guards come in several lengths. For maximum protection,

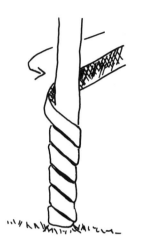

Figure 94. *Installing a plastic mouse guard*

select one that will reach from the soil surface to the bottom of the first scaffold. Leave the guard in place until the tree trunk has expanded and begun to unravel the guard. At this time the trunk is probably of sufficient diameter to withstand a small amount of mouse feeding. It has also grown less tender and tasty to the mice by this age.

Rabbit feeding will also be controlled reasonably well by mouse guards during the fruit tree's early years. If you have a severe rabbit problem, fashion a wider diameter guard with a piece of chicken wire to serve as permanent protection once the mouse guard has been outgrown.

Many backyard gardeners mulch their fruit trees and in so doing create the perfect habitat for mice. One easy preventive step is to keep all mulch back away from the trunk six to twelve inches. (This is good strategy for disease prevention too.) In the fall of the year, rake the mulch back even further to discourage rodents from setting up their winter home near the tree's trunk.

Raccoons and Porcupines

Raccoons commonly cause damage by breaking many branches in their pursuit of ripe fruit. Very often the only sign of raccoon damage in peaches are the pits left under each tree when the feast is done. Wayward porcupines can be an occasional pest in fruit tree plantings. They will commonly climb into the tree several feet and gnaw off branches giving the tree the appearance of a bad hair cut. Control consists of trapping the animal in a live trap and removing it to more hospitable territory. Shooting it is also an option, but is not well received in a suburban environment.

Section V.

Harvest

18. Harvest and Storage

Once your home orchard is growing and ready to produce its first crop the question "how do I know when the fruit is ready to pick?" will surely enter your mind. Fortunately there are several ways to determine this. Experience will become your best guide.

When to Harvest for Fresh Eating

One way to determine a ballpark date for harvesting your fruit is by being aware of how many days have elapsed since full bloom. Experience has shown that the time between full bloom and harvest is quite constant from year to year regardless of the weather. Apples and pears, in particular, are quite dependable in the elapsed time needed to develop proper harvest maturity. Table 29 gives the normal ripening time needed for some of the more common apple varieties. More information on pears can be found in Chapter 5.

Table 29 - Days from Bloom to Harvest (Apple)

Variety	Days from bloom to harvest
Cortland	125-130
Golden Delicious	140-145
Granny Smith	180-200
Gravenstein	110-115
Grimes Golden	140-145
Jonathan	140-145
McIntosh	125-130
Northern Spy	145-155
Rhode Island Greening	135-145
Wealthy	120-125
Winesap	160-170
Yellow Transparent	70-95

Although each variety has its own requirements these figures can be used as a starting point to aid in applying other maturity indicators.

For the backyard gardener desiring to eat most of the harvest at picking time, harvest readiness can be judged by observing changes in skin color, seed color, or simply picking a few fruit to see how easily they come loose from the tree. Taste testing, for most varieties, is of course the ultimate indicator if you plan to use the fruit soon rather than storing it for an extended time.

Most fruits will show a change in the green undercolor (or ground color) of their skin as they ripen. The dark grass green will often change to a lighter yellow green and then to the respective ripe fruit's normal skin color - usually red, purple, or yellow. This change in tone of the green undercolor is a reliable indicator of approaching ripeness. Be careful, however; don't rely heavily on the amount of red skin color. It is effected quite a bit by the amount of sunlight received just prior to harvest time. Seasons in which cool nights are accompanied by warm sunny days will encourage red color development. For apples, cloudy daytime conditions or warm nights often result in fruit that is a dull red and not extensively colored. A change in green ground color will occur regardless of the weather.

A change in seed color is another signal of approaching maturity in pome fruit (apples and pears). When the fruit is immature, the outside of the seed is a pale creamy white. As the fruit nears ripeness, the seed coating will turn first light and then dark brown in color. By cutting open a few sample fruit, you can easily check how rapidly picking time is approaching.

Some fruit, most noticeably the McIntosh apple and its related types, often show maturity by dropping a number of fruit from the tree if they are not picked soon enough. Many of the earlier maturing summer apple varieties ripen unevenly and will do the same. Often by picking a few random fruit, you will feel how easily they come loose from the tree, signalling maturity.

Most fruit growers come to rely on a combination of these factors to decide when to start picking. What you intend to do with your fruit will also play a part in your timing. Fruit that you plan to eat fresh right away can be left on the tree until it is fully ripe. Likewise, fruit that is meant for storage needs to be picked slightly underripe so that it does not become over mature in storage.

Picking Technique

The thought of picking a fruit seems so elementary that you may question why one would even think to discuss it, but actually there is more to it than first meets the eye. How the fruit is picked can have an effect on both the quality of the fruit and in some cases the amount of fruit produced by the tree the following season.

First, you should avoid bruising or otherwise physically damaging the fruit you are about to pick. This can be quite important since fruit that is bruised will age more quickly and lose quality during storage.

Second, trees that bear fruit primarily on spurs, as apples, pears, and plums do, rely on the spurs for future production of fruit buds. If the spurs are broken off in the picking process, eventually the productive capacity of the tree may be reduced.

For most fruit, using a "twist and lift" technique when you pick will make picking easier while at the same time insuring that the fruit and tree remain undamaged. Rather than just grasping the fruit in your hand and pulling toward you, cradle the fruit gently but firmly in the palm of your hand. Support the stem between your thumb and index finger. Then slowly rotate your whole hand while lifting the fruit upward toward the spur. Assuming the fruit is ripe, it should release easily and cleanly from the tree. With a little bit of practice, you should

Figure 95. *"Twist & lift" technique for easy fruit picking*

soon be able to master this technique and picking will become an effortless fluid motion.

When placing fruit in your basket or other picking container, be sure to do so gently. Many people do not realize that even an apparently hard pear or apple can bruise easily. One of the most common ways that fruit becomes bruised is from being dropped into the picking basket or handled roughly when sorted.

Fruit will generally retain its quality better if it is picked with the stem still attached to the fruit. In most cases, if the fruit is picked with the technique just discussed, the stem will remain attached. Notable exceptions are peaches, nectarines and plums. Here the stems naturally release from the fruit when ripe and can do more damage by puncturing adjacent fruits' skin than can be gained by keeping the stem on.

Table 30 - Proper Fruit Storage Conditions

At some point, most backyard orchardists experience the mixed blessing of a bountiful harvest. In order to make use of the plentiful supply, some fruit may need to be preserved or stored for later use. Naturally you will want to store it so that it retains its "just picked" quality until you can use it. One of the most important factors in keeping fruit fresh is temperature. How fast fruit is cooled to storage temperature is also important. Ideally you should store fruit under refrigeration, with sufficient humidity to prevent drying and shriveling. The chart below shows optimum storage temperatures for a number of fruits.

Proper Storage Conditions

	Temperature (°F)	Humidity (%)	Storage Time
Apple	30-40	90	2-6 months
Apricot	32	90	up to 2 weeks
Peach/Nectarine			
Clingstone	32	90	2 weeks
Freestone	32	90	up to 1 month
Pear	30	90-95	2-7 months
Plum	32	90-95	up to 1 month
Sweet Cherry	32	90-95	2-3 weeks
Tart Cherry	32	90-95	1 week

Until not so long ago, the root cellar was used to store all sorts of produce. If you have access to one, it works as well as ever. For most of today's home gardeners, a spare refrigerator in the basement or garage often takes the place of the root cellar. Both provide an important ingredient for proper storage - cool, consistent temperature. Lacking an extra refrigerator, many people in locations with cool winter temperatures have fared reasonably well, wrapping pears and apples in newspaper and storing them in a plastic bag inside a large plastic trash can. Kept in a cool location, this can substitute for the old fashioned root cellar. If you use this method be sure though to protect the fruit from actual freezing temperatures.

Two other considerations are important in storing fruit from the backyard orchard; see that the fruit is refrigerated quickly after picking and stored so that it will not dry out and shrivel. As soon as possible after picking, refrigerate your fruit. It comes as a surprise to many people that a fruit picked in 80 or 90° F heat may take several days to cool to 35° F at its core, even when quickly refrigerated. The longer the fruit is warm the faster it will ripen and age. For each extra day that a harvested apple spends at 70° F, it can loose a week to ten days of its storage life.

To keep your fruit moist and juicy, storing it in a plastic bag or covered container helps. For apples and pears, poke a few holes in the bag and, with wet finger tips, you may add a *light* sprinkle of water to the bag. Do not add water to stone fruits in storage though; it can quickly encourage the growth of rots and molds in their very soft flesh.

Following proper storage, certain fruit benefit from additional ripening at higher temperatures. Peaches, nectarines, apricots, plums, and pears ripen best when exposed to 65° F for two to three days. Apples and cherries are best when used directly from refrigeration.

Section VI.

Resources & References

Appendix 1

How to Revive an Ailing Tree

Reviving a distressed tree is often like solving a puzzle. The answers come together piece by piece. Some answers are much more obvious than others. For example, if your developing fruit has marks on it that look clearly like something was chewing on it, your problem is very likely to be an insect. You don't even need to refer to this chart, you can go directly to the chapters on insect identification and control. Since insect and disease related problems each take up a chapter in themselves, the obvious problems have not been included in this section. Instead this chart will concentrate on solving problems whose cause is not always readily apparent.

If you have read this book through completely, you most likely have come across the answers you need, but right now you may not be sure where you saw them. By following the questions in this chart you should be directed once again to the solution.

Most of the problems that occur with fruit trees are first observable at a particular and predictable time of the season. So, to give you an easy starting point in your search for a solution, problems are listed under the season in which they are most likely to be observed. Start by looking under the appropriate seasonal heading. If your problem is not present there, go to the previous season. Often problems develop long before symptoms actually become visible. Variations in climate throughout the United States may also account for some differences in timing.

Where several factors could be causing a problem, they are listed as "See also:....", followed by a page number in parenthesis for short explanations. For more extensive information you are referred to a chapter.

Winter

??? The bark looks shrivelled and dry. The tree has no leaves.
The tree may be experiencing windburn. Wait until spring to see if the tree grows properly.

??? Leaves began to grow and have turned brown/black and died.
The tree likely started to grow during a warm spell and then was damaged by subsequent cold. See also: chilling requirements (150-151), excess nitrogen fertilizer (123)

??? Something has eaten the bark around the trunk.
This is most likely rodent feeding. See (201). Paint with grafting emulsion. If the trunk has been girdled all the way around, a bridge graft may be the only remote solution.

??? The trunk of my tree just split wide open.
This is likely sunscald or "southwest injury". See (43)

??? The melting snow has broken some of the limbs on my tree. What do I do.
The broken limbs should be removed. You can prune them off now or in early spring before growth starts.

Spring

??? Most of the branches on my tree are leafing out, but some don't have any leaves and look like they might be dead.
They probably are dead. If your tree is old, this is probably just natural and normal. If the trees are young, they may have suffered cold injury. See Chapter 11 for more on pruning.

??? The tree does not have blossoms or leaves, but other trees in my yard are starting to grow.
If the tree is a very late blooming variety, such as York, Rome, Greening, or Northern Spy apple this could be normal. Otherwise your tree may have suffered winter kill and will need to be replaced with a variety that is better suited to your winter climate. See (24)

??? **The tree has blossoms, but no leaves.**

Most fruit will bloom before their leaf buds are fully open. As long as the leaves are beginning to appear, this is normal.

??? **Last year my tree had lots of fruit. This year it doesn't even have any bloom.**

A number of fruit varieties can develop a habit of bearing every other year if allowed to crop too heavily. See biennial bearing (156)

??? **The blossoms started to open, but have now turned brown, wilted and stopped growing.**

A severe spring cold snap is probably to blame.

Open blossoms of apple and pear can be quite susceptible to fireblight under certain weather conditions. See (182 & 197)

??? **All the petals are falling off the blossoms.**

This is normal after the flower has been in bloom for five to seven days. See developmental stages of fruit (48 & 81)

??? **All the branches of my new tree have been eaten off by something. What do I do?**
??? **The ends of new shoots on my trees seem to have been broken or eaten off.**

Deer feeding is probably the cause. Prune the ragged tips off to a remaining shoot or bud and wait to see if new shoots develop to replace the damaged ones. See (200) for ways to prevent further damage.

??? **The leaves seem to be wound up tightly and are not unrolling & growing as I expected.**

This could be leaf roller damage. See insect identification (166) and pest control (195)

Summer

??? The fruit seemed to be growing and now all of a sudden much or all of it is falling off the tree.

June drop is probably occurring. A certain amount of fruit drop is normal, see (156).

If you sprayed apples with Sevin insecticide, this may be causing fruit thinning. Use some other insect control remedy and remember not to spray apples with Sevin in the first thirty days following bloom.

Poor fertilization may be another cause, see (155-156).

Curculio and codling moth also can cause fruit drop. See Chapter 14.

??? My tree seemed to be growing and all of a sudden several branches (or the whole tree) started to wilt and die.

Did you have an unusually cold winter or a sudden cold snap that followed a warm spell?

YES: Your tree may have suffered winter injury, but was able to live for a short time on its stored carbohydrates. With a pocket knife, cut a slice of bark down to the cambium layer. If the cambium is brown, cold injury has occurred. If only certain limbs are dead or weak, prune them back to healthy new wood and hopefully the tree can be revived.

Is your tree only several years old?

YES: You may have a graft incompatibility problem. See the answer to the next question.

??? The tree has suddenly broken off at or near ground level.

You very likely have a graft incompatibility problem. Consult an experienced nursery operator for a compatible scion/rootstock combination.

??? The leaves are turning light yellow or have purple veins or margins.

Symptoms point to a nutrient deficiency, see Chapter 10.

??? **My trees leaves have brown, burned looking edges that are drying out.**

You may have put too much fertilizer down too close to the tree. Especially when the tree is young, its roots are sensitive to fertilizer burn. See page (38) and be sure to spread most of your fertilizer in the drip line area, not right next to the trunk.

??? **The tree leaves are turning a brownish red, almost bronze color.**

Check for mites. Heavy infestations can cause leaf "bronzing". See (168) for identification and control (174 & 196).

This could be a nutrient deficiency. See (122).

??? **The leaves are getting a thick wavy appearance and don't look normal.**

Is this a peach tree? See peach leaf curl (181 & 197)

Is this a plum or pear tree? Consult your local extension specialist, this may be a minor occasional disease or insect problem that is limited to your local area.

??? **The tree leaves are turning yellow and falling off.**

The likely cause is a fungus disease or virus; cherry leaf spot, peach leaf curl, bacterial spot, or X-disease. See Chapter 15. Consult an extension agent for diseases specific to your area.

If you are in a very dry area, have planted on a very sandy soil, or experiencing an abnormally dry summer, the tree could be suffering severe drought stress. Water at once and see (123).

??? **Leaves or whole branches of my tree are turning dark brown, almost black.**

Sounds like fireblight, most common in pears and certain apple varieties, see (182)

??? **My fruit has spots on it.**

Most likely cause is a fungus disease, apple or pear scab (177), bacterial spot (183). See Chapter 15 for other diseases

??? My fruit is all knarled or misshapen.
This could be caused by stony pit virus on pears (185), severe apple scab infection (177), curculio stings (165), or catfacing insects (171).

??? My fruit is getting this grey mold on the outside.
Is your fruit a peach, cherry, apricot or plum?
YES: Brown rot fungus is probably the cause. See (178)

??? The leaves are all tattered and chewed up.
Insect feeding is likely to blame, see Chapter 14. Unusually high winds may tatter the leaf edges on occasion too. Carefully maintain the tree for the remainder of the season and be diligent about providing protection against fireblight (197).

??? My fruit seems to have worms in it.
Undoubtedly an insect problem, most likely codling moth, possibly oriental fruit moth, curculio, apple maggot or cherry fruit fly. See Chapter 14 & 16.

??? My peaches are really big this year, but some of them seem to have almost split open and in others the pit has broken into several pieces. What's wrong.
It sounds like you are experiencing "split pits", a disorder that occurs in some years when moisture supply is uneven. See (113).

Fall

??? My fruit seemed almost ready to pick, but still didn't taste sweet; then all of a sudden much of it dropped off the tree.
Some varieties of apple, especially McIntosh have a tendency to drop at maturity, see (205-206)

??? My fruit isn't turning red the way it usually does.
Certain weather conditions discourage extensive red color formation. Cloudy days and warm night temperatures could be the culprit. The fruit may not look as beautiful, but if all other maturity indicators are present, you should pick the fruit and rest assured it will still be just as good to eat. See Chapter 18.

??? The fruit doesn't taste sweet.

Consult Chapter 18 to be sure the fruit is indeed mature. Some weather conditions, such as a cold rainy season may produce fruit that is not as flavorful as fruit produced in a warm sunny season.

??? My pears looked so perfect, but they're rotten inside.

Domestic pears must be harvested immature and ripened off the tree to avoid internal breakdown, see (74)

??? My apples have corky brown spots on the skin that show up just before or after harvest.

This is most likely bitter pit, a physiological disorder often caused by calcium deficiency. See (122 & 120).

??? My tree seems to be growing when I think it should be slowing down for winter.

Excess nitrogen could be stimulating growth, see (123)

Late season pruning also encourages continued growth, see (130)

Any Time of Year

??? My tree has been growing for years and still doesn't have any fruit.

Fruit production can be influenced by many things. First the type and age of your tree, see (154).

Is your tree producing flowers?

NO: It may be receiving excess nitrogen, see (123)

YES: It may require cross pollination by another tree, see (151-156)

Do you prune your tree heavily?

NO: It may need more sunlight, see (117 & 130)

YES: You may be overstimulating vegetative growth, see (130-131)

Do you live in an area that receives frequent spring frosts?

YES: Your blossoms may be damaged by the frost. Try growing a variety that blooms after the last frosts in your area.

??? My tree has been growing for years and it's still only 5 feet tall.
Your tree is probably grafted on a very dwarfing rootstock. See the rootstock section in the chapter on your particular fruit.

Are the leaves a healthy dark green?
YES: Other reasons given below are likely causes.
NO: A nutrient deficiency may be the cause. See Chapter 10.

Is your soil a very heavy clay?
YES: You may have selected an unsuitable rootstock for your soil conditions. Most roots have trouble growing in very heavy, wet soils. See Chapter 2. In order to loosen up the soil structure, mulch the ground around your tree annually. This will not solve the problem quickly or completely but may be all you can do short of replacing the tree with a more suitable one.

Repeated deer feeding can stunt trees. See Chapter 17 for measures to prevent further damage.

If your tree is planted in a container, it could be rootbound. Is it showing other signs of poor growth; short side shoots, pale leaves, poor quality fruit. See Chapter 12.

??? The fruit on my tree is always small.
You may need to thin your fruit after bloom. See Chapter 13.

If you have not pruned your tree recently, pruning may help improve the fruit size. See Chapter 11.

Some varieties are naturally smaller in size than others. If the tree is healthy and being cared for as suggested in the preceding chapters, enjoy your fruit's flavor, eat two, and don't worry too much about its size.

??? The fruit on my tree isn't anything like what I expected from the catalog description.
You may have received an incorrectly labeled tree or the graft may not have survived and what you have is actually the fruit of the rootstock. See the question below and consult the nursery from which you purchased the tree.

??? I thought I bought a dwarf tree but it is growing much bigger than I expected.

Make sure that you planted the graft union above ground. If you failed to do so the scion has probably rooted and you will have a tree closer to standard size. See (40) or the previous question.

??? My tree has blown over after a big storm. What do I do.

This may have happened for several reasons: your tree is growing on a rootstock that naturally has poor anchorage or brittle roots; the rootstock was not well suited to your soil conditions; or the storm was just unusually bad. If the tree still has a portion of its roots in the ground, you could try propping it up and replanting. Realize that it will need extra care and the support of a stake to survive. It probably will not be able to support a heavy fruit load in the future and you will have to decide if it is worth saving. If you choose to replant, read Chapter 2 and the section on rootstocks in the respective fruit chapter.

Appendix 2

The following almanac is intended to give the gardener a representative idea of the maintenance activities that might occur in caring for a fruit tree throughout the four seasons. This time frame is typical to what you can expect in the upper midwest and New England. In the warmer climates of the west coast and the south, the activities will be similar, but will occur several weeks earlier.

Annual Almanac

January

Apple: Prune older trees and most winter hardy varieties.

Pear: Prune mature trees.
            Prune fireblight from all trees.

February

Collect budwood if you plan to do any spring propagating.
Inspect tree trunks for mouse and rabbit feeding.

Apple: Prune mature trees if not done earlier.

March

Cherry: Prune mature trees at the beginning of the month, young trees toward the end of the month.

Plum: Prune out and burn black knot fungus.

April

Spread new mulch under trees.
Start acclimating container grown plants.
Tune up equipment, see that your sprayer is working properly.
A good time for spring grafting and propagating new trees.
Apply fertilizer to all but peach trees.
Plant new trees.

Apple: Bordeaux spray for fireblight on susceptible varieties.

Pear: Bordeaux spray for fireblight on susceptible varieties.

Peach: Apply peach leaf curl spray if not done last fall.

May

Apple: Apply dormant oil spray early in the month.
           Start checking for leafrollers, green fruit worms, curculio,
           and tarnished plant bug by the end of the month.
           Expect bloom to start toward the end of the month.

Pear: Apply pear psylla spray.
           Check for same insects as apples.
           Expect bloom from middle to the end of the month.

Cherry: Bloom expected by mid-month.
           Possible need for a brown rot spray if wet weather occurs
           during or after bloom.

Apricot: Bloom early in the month.
           Prune after bloom is done.
           May need to apply brown rot sprays if wet weather occurs
           during or after bloom.

Peach: Bloom occurs early to mid-month.
           Prune after bloom.
           Apply calcium nitrate fertilizer after bloom, based on
           anticipated crop size.

Plum: Spray for black knot.
           Bloom.

224 the Backyard Orchardist

June

Apple: Monitor insects and scab development conditions.
           June drop of excess fruit should occur mid month.
           Begin hand thinning after June drop, if needed.
           Soil test and begin preparing site for next year's planting.

Pear: Continue fireblight control as conditions require.
           Hand thin fruit.

Cherry: Start watching for cherry fruit fly.

Apricot: Hand thin as needed.

Peach: Begin hand thinning as needed.
           Apply second fertilizer application as needed.

July

Apple: Start watching for apple maggot and mite build-up.
           Summer prune bearing trees.
           Train young trees.

Cherry: Harvest fruit.

Apricot: Harvest fruit toward end of month.

August

Apple: Summer varieties ready for harvest.
           Good time to propagate trees.
           Summer prune bearing trees, if not done earlier.

Pear: Harvest early varieties.

Apricot: Harvest fruit.

Peach: Harvest fruit.

Plum: Early varieties ready for harvest.

September

Apple: Harvest early fall varieties.

Pear: Harvest Bartlett and other early fall varieties.

Peach: Harvest late varieties.

Plum: Harvest fall varieties.

October

Apple: Harvest.
          Make cider.

Pear: Harvest late varieties.

November

Pull mulch back from trunks.
Winterize container plants.
Winterize equipment.

Peach: Apply peach leaf curl spray now or early next spring.

December

Order new trees for planting next spring.
Enjoy fruit you have stored and preserved.
Catch up on reading your gardening magazines.
Give a gift certificate for a fruit tree and a copy of *the Backyard Orchardist* as a holiday gift.

Resources

Since many of the more uncommon fruit varieties are often difficult to find, this section of nurseries has been included to help you find just what you are looking for should it be unavailable locally. All of the nurseries below were surveyed prior to publication of this book. Comments are based on survey responses and/or review of their catalog. A listing does not necessarily constitute an endorsement of the business or the quality of their products. In the interest of keeping current, the publisher welcomes information about your experience with the businesses listed below or about other mail order nurseries that should be included in future editions.

Vernon Barnes & Sons
PO Box 250
McMinnville, TN 37110
615-668-8576
Limited selection of dwarf trees, greater selection on standard rootstock

Bear Creek Nursery
PO Box 411
Northport, WA 99157
Descriptive catalog with a wide selection of varieties, rootstock, scionwood, tools and other fruit related supplies

Bluebird Orchard & Nursery
429 Randall St.
Coopersville, MI 49404
Listing of antique apple varieties. Scionwood

Bottoms Nursery
Rt. 1, Box 281
Concord, GA 30206
706-495-5661

Boyer Nurseries
405 Boyer Nursery Rd.
Biglerville, PA 17307
717-677-8558
Price list but, no descriptions

Buckley Nursery &
Garden Center
646 N River Rd.
Buckley, WA 98321
206-829-1811
Price list with brief descriptions

Burford Brothers
Rt. 1 - Nursery
Monroe, VA 24574
804-929-4950
Excellent selection of stock order and custom propagated antique apple varieties. Also rootstock and orchard supplies

Burnt Ridge Nursery
432 Burnt Ridge Rd.
Onalaska, WA 98570
206-985-2873
Good selection of Asian pears, disease resistant peaches & apples. Most fruit on newer dwarf rootstocks

W. Atlee Burpee Co.
300 Park Ave.
Warminster, PA 18991
215-674-4915
Well known garden catalog; includes a few tree fruit

C & O Nursery Co.
PO Box 116
Wenatchee, WA 98807
509-662-7164

Chestnut Hill Nursery
RR 1, Box 341
Alachua, FL 31615
800-669-2067
Southern pears, nuts, & more

Christian Homesteading
Movement
Oxford, NY 13830 A
Scionwood - send SASE. Educational workshops

Classical Fruits
8831 AL Hwy. 157
Moulton, AL 35650
205-974-8813
Descriptive catalog with wide selection of fruit, including numbered disease resistant apples. Scionwood

Cloud Mountain Farm
6906 Goodwin Rd.
Everson, WA 98247
206-966-5859
Antique varieties

Cumberland Valley Nurseries
PO Box 471
McMinnville, TN 37110
615-668-4153
Extensive selection of peaches including chill requirements. Smaller number of other tree fruits

Edible Landscaping
PO Box 77
Afton, VA 22920
800-524-4156
Small selection of primarily southern fruit

Exotica Rare Fruit Nursery
PO Box 160
(2508 B. E. Vista Way)
Vista, CA 92085
619-724-9093
*Descriptive list of Asian pears,
low chill apples, exotic fruit*

Farmer Seed and Nursery Co.
818 NW 4th St,
Faribault, MN 55021
507-334-1651
*Small selection, mostly hardy
varieties for northern climates*

Fedco Trees
Box 340
Palermo, ME 04354
207-872-9093
*Descriptive listing with good
selection on hardy varieties.
Books & supplies*

Fowler Garden Center &
Nurseries, Inc.
525 Fowler Rd.
Newcastle, CA 95658
916-645-8191
*Price list with good selection of
varieties, mostly suited to West
Coast. A few genetic dwarfs.
Home Orchard Guide for sale
gives variety descriptions*

Greenmantle Nursery
3010 Ettersburg Rd.
Gaberville, CA 95542
*Extensive apple & pear selection:
smaller, but good selection of
locally adapted fruit*

Gurney's Seed & Nursery Co.
110 Capitol St.
Yankton, SD 57079
605-665-1671

Harmony Farm Supply
PO Box 450
Graton, CA 95444
707-823-9125
707-823-1739 FAX
*Nice catalog of tools, pest con-
trol supplies, books $2.00. Small
selection of fruit trees, must be
picked up at the nursery*

Henry Field Seed & Nursery
Company
415 N. Burnett
Shenandoah, IA 51602
605-665-4491
*General garden catalog, includes
fruit; small selection of hardy &
disease resistant varieties*

Hidden Springs Nursery
Rt 14, Hidden Springs Ln.
Cookeville, TN 38501
615-268-9889
*Small catalog with eclectic collec-
tion of fruit that does well in
Tennessee, a few disease resistant apple on M106 & M111
and a few pears*

Ison's Nursery & Vineyards
Brooks, GA 30205
800-733-0324
*Low chill apples and other fruit
suited to the south*

Johnson Nursery
Rt. 5, Box 29J
Ellijay, GA 30540
706-276-3187
Antique & older varieties

Lawson's Nursery
Rt 1, Box 472
Ball Ground, GA 30107
706-893-2141
Antique varieties-some on newer dwarf rootstock; Colt, GM-61, P-22. Container plants on M27

Henry Leuthardt Nursery
Montauk Hwy., Box 666-BOG
E. Moriches, Long Island, NY 11940
516-878-1387
Antique varieties. Espalier trees.

Long Hungry Creek Nursery
Box 163
Rd Boiling Springs, TN 37150
Specializes in disease resistant apples, on M111 rootstock

Living Tree Center
PO Box 10082
Berkley, CA 94709
510-420-1440
Descriptive catalog of historic varieties. Scionwood

Mellinger's
FGB West Range Rd.
North Lima, OH 44452
216-549-9861

J E Miller Nurseries, Inc.
5060 W Lake Rd.
Canadaigua, NY 14424
716-396-2647
800-836-9630
Descriptive catalog of tree fruit & berries, includes a few Asian pears. Supplies

National Arbor Day Foundation
100 Arbor Ave.
Nebraska City, NE 68410
Selection of old stand-by varieties

Northwind Nursery
& Orchards
7910 335th Ave., NW
Princeton, MN 55371
612-389-4920
Hardy northern fruit varieties, supplies, books, workshops

Northwoods Nursery
27635 S. Oglesby Rd
Molalla, OR 97013
503-226-5432
Informative catalog selection includes disease resistant, antique, asian pear, pluot & aprium varieties; some grafted on newer dwarf rootstock.

Oregon Exotic Rare Fruit Nursery
1065 Messenger Rd.
Grants Pass, OR 97527
503-846-7578
A few conventional tree fruit, many unusual fruits. Catalog $2

Pony Creek Nursery
PO Box 16 Nursery Ln.
Tilleada, WI 54978
715-787-3889
Catalog with small selection of mostly hardy northern varieties

Raintree Nursery
391 Butts Rd.
Morton, WA 98356
206-496-6400
Informative catalog with excellent selection of all the types of fruit discussed in this book & more. Rootstock & supplies

Raven Island Nursery
Waldron Island, WA 98297
Scions & budwood of antique varieties

Rocky Meadow
Orchard & Nursery
360 Rocky Meadow NW
New Salisbury, IN 47161
812/347-2213
Good selection of varieties with emphasis on flavor & disease resistance, on wide choice of newer dwarf rootstocks

Sonoma Antique Apple
Nursery
4395 Westside Rd.
Healdsburg, CA 95448
707-433-6420
Apple & pear scionwood, container stock, espalier trees. Wide selection. Descriptive catalog $2.00 refunded with order

Southmeadow Fruit Gardens
Box SM
Lakeside, MI 49116
616-469-2865
Excellent selection of "choice & unusual fruit for the connoisseur", propagated on wide choice of rootstocks. Also rootstock for sale. Illustrated 112 page catalog with descriptions & history available for $9.00

catalog

Stark Brothers
PO Box 10
Louisiana, MO 63353
800-325-4180
General selection of varieties adapted to a range of conditions. City Orchards catalog for the suburban gardener

Van Well Nursery
PO Box 1339
Wenatchee, WA 98807
509-663-8189
Nursery serving the commercial grower that also sells in smaller lots. Descriptive catalog with good selection of varieties & rootstock

M. Worley Nursery
98 Braggtown Rd.
York Springs, PA 17372
717-528-4519
Price list only, no descriptions. Good selection of varieties, especially peach

Publications

The following is a random list of publications the author has found helpful in locating additional information, resources, or rare plant material that other fruit gardeners may also enjoy.

Fruit Nut & Berry Inventory
Seed Savers Exchange, 3076 North Winn Rd., Decorah, IA 52101
A descriptive index of over 5800 fruit and nut varieties with cross referencing on which of over 300 mail order nurseries carries each variety.

Garden Literature: An Index to Periodical Articles & Book Reviews
Garden Literature Press, 398 Columbus Ave, Ste. 181, Boston, MA 02116-6008
Indexes over 100 English-language periodicals; some horticultural, some general interest; that include garden topics. Issued quarterly with fourth quarter as an annual cumulation.

Gardener's Index
Compudex Press, PO Box 27041, Kansas City, MO 64110-7041
Annual comprehensive subject index of 6 major U.S. gardening magazines (Am. Horticulturist, Fine Gardening, Flower & Garden, Horticulture, National Gardening, and Organic Gardening).

Gardeners Source Guide
PO Box 206, Gowanda, NY 14070-0206
Listing of mail order nurseries, including fruit, that send free catalogs as well as associations, societies, and clubs.

Gardening By Mail by Barbara Barton
A comprehensive listing of plant sources, including many fruit tree and berry sources. Available at most bookstores.

Glossary of Terms for the Home Gardener by Robert Gough, PhD
Haworth Press, 10 Alice St., Binghamton, NY 13904
A handy pocket glossary of briefly defining many common gardening terms. Not limited to fruit production, this is a ready reference for any gardener striving to better understand gardening and its

Further Reading

Magazines
Common Sense Pest Control Quarterly and The IPM Practioner.
Bio-Integral Resource Center, PO Box 7417, Berkley, CA 94707.
Country Journal. 4 High Ridge Park, Stamford, CT 06905.
The Edible Landscape. 115 Vandewater St., Providence, RI 02908.
Great Lakes Fruit Grower News. PO Box 128, Sparta, MI 49345
Horticluture. 98 N. Washington St., Boston, MA 02114-1913.
HortIdeas. 460 Black Lick Rd., Gravel Switch, KY 40328.
National Gardening. 180 Flynn Ave., Burlington, VT 05401.
Organic Gardening. 152-180 E.Minor St., Emmaus, PA 18098.

Books
General Fruit Growing
Creasy, Rosalind. *Organic Gardener's Edible Plants.* Van Patten
Publishing, Portland, OR. 1993.
Gordon, Donald. *Growing Fruit in the Upper Midwest.* University of
Minnesota Press, Minneapolis, MN. 1991.
McEachern, George. *Growing Fruits, Berries & Nuts in the South.*
Gulf Publishing Co. Houston, TX. 1989.
Reich, Lee. *Uncommon Fruits Worthy of Attention.* Addison-Wesley
Publishing Company, Inc., Reading, MA. 1991
Walheim, Lance and Robert L. Stebbins. *Western Fruit, Berries, and
Nuts:How to Select, Grow, and Enjoy.* H.P. Books, Tuscon, AZ, 1981.

Soil Fertility/Composting/Mulching
. *Backyard Composting.* Harmonious Technologies, Ojai, CA. 1992.
Solomon, Steve. *Organic Gardener's Composting.* Van Patten
Publishing, Portland, OR. 1993.
Willis, Harold. *The Coming Revolution in Agriculture.* Harold L.
Willis, Wisconsin Dells, WI. 1985.
Lee, Andrew W. *Chicken Tractor: The Gardener's Guide to Happy
Hens and Healthy Soil.* Good Earth Publications, Shelburne, VT.
1994.

Pest & Disease Identification and Control
Ellis, Barbara W. and Fern Marshal Bradley. *The Organic Gardener's
Handbook of Natural Insect and Disease Controls.* Rodale Press,
Emmaus, PA. 1992.

Organizations & Miscellaneous Resources

Applesource
Rt 1, Chapin, IL 62628   Phone: 217-245-7589
Applesource sells mail order apple samplers for taste testing. A great way to try a variety before you plant a tree! Their catalog is a useful reference about flavor, keeping quality, and harvest time of many of the best tasting apple varieties, both old and new.

Home Orchard Society
PO Box 230192, Tigard, OR 97281-0192
A non-profit educational organization assisting new and experienced fruit growers with fruit bearing trees, shrubs, and plants in the home landscape. Annual membership of $10.00 includes quarterly POME NEWS journal, branch and chapter events, and use of media library. Major events include a spring scionwood exchange and rootstock sale, summer orchard tours, and autumn All About Fruit Show.

North American Fruit Explorers
Rt. 1, Box 94, Chapin, IL 62628
NAFEX, for short, is a network of over 3000 individuals interested in discovering, cultivating and furthering knowledge of superior fruit and nut varieties. Membership is $ 8.00 for one year or $ 15.00 for two. Membership benefits include the quarterly journal, POMONA, an 80 page collection of articles written by members; a lending library book list and borrowing privileges; a propagating stock exchange; special interest group participation; and an annual meeting.

Seed Savers Exchange
3076 North Winn Rd., Decorah, IA 52101
Seed Savers Exchange is a non-profit organization dedicated to saving heirloom and endangered varieties of food plants. Their Heritage Orchard, planted in 1989 and open to the public during the summer, includes over 700 varieties of apples documented to have grown in North America previous to the early 1900s. Limited quantities of scionwood will be available starting in 1995. A small collection of extremely hardy grapes has also been planted. Publish the Fruit, Nut, and Berry inventory; an extensive list of varitiey descriptions and nursery sources. Annual membership is $25.00 or send $1.00 (to help defray costs) for a full color brochure further outlining member benefits and projects.

234 the Backyard Orchardist

Worcester County Horticultural Society
30 Tower Hill Rd., Boylston, MA 01505 Phone:508-869-6111
The society has been furthering the practice of horticulture for over 150 years through many activities, including development of the S. Lothrop Davenport Preservation Orchard, a collection of over 115 antique apple varieties. Sales of scionwood from the orchard helps fund society projects.

Cooperative Extension Service (CES) and Master Gardener Program

As a cooperative effort between federal, state, and local government, the extension service disseminates information based on the research conducted by the land grant colleges or universities in each state. The extension service has a number of services for the home gardener including informational bulletins, seminars, soil testing services, and the advice of a horticultural or agricultural agent. Most of these services are moderately priced and some are free of charge.

Most states have an extension office in each county. You may find the office listed under any of the following headings in your local phone book - agricultural agent; cooperative extension service; County, extension service; Federal or U.S. government, extension service; Land grant university name, extension service.

If you need the services of your extension office, most state programs are set up in such a way that you contact your local office. If they are unable to immediately answer your question, they will contact a specialist at the state land grant college and obtain the answer for you or refer you to that specialist.

The Master Gardener Program, an out-growth of the cooperative extension service's educational programing has two functions:

1. as an educational program providing current horticultural information to active gardeners, and

2. as a service program of volunteer activities provided by Master Gardeners.

Master Gardeners receive approximately 40 hours of classroom and field training in horticulture and participate in an equal amount of time providing community garden services. Approximately half of this time is spent working with cooperative extension in answering "garden hotline" questions or developing educational materials. The balance of volunteer time is spent on community beautification projects, providing horticultural therapy, staffing informational booths, or other community gardening activities.

Resources 235

Appropriate Technology Transfer for Rural Areas (ATTRA)
PO Box 3657, Fayetteville, AR 72702 Phone: 800-346-9140
This US Fish and Wildlife Service program provides free information on low input/sustainable agriculture. Several reports are available on raising fruit as well as more general soil fertility management and pest control.

Office of Small Scale Agriculture
USDA/CSRS, OSSA, Ste. 328-A, Aerospace Center
Washington, DC 20250-2200 Phone: 202-401-1805
Geared to providing a wide array of information to the small scale farm and specialty crop grower, this USDA office provides fact sheets and reports on a number of topics related to fruit. A free newsletter covering numerous agricultural/horticultural topics is also available.

National Agricultural Library, USDA, Room 111, Beltsville, MD 20705. Phone 301-504-5755.
The library houses primarily technical publications on all aspects of agriculture. Items may be requested through interlibrary loan from public and university libraries. It also offers AGRICOLA, an on-line computer data base, and an electronic bulletin board.

Glossary

Acid soil. Soil with a pH less than 7.0. An acid soil is usually low in lime. Also caused by application of high amounts of fertilizer. Most often found in rainy climates.

Acidity. see Acid soil.

Alkaline soil. Soil with a pH greater than 7.0. Most often found in arid or desert climates. Can be modified by the addition of sulphur.

Alkalinity. see Alkaline soil.

Anther. (Male) pollen bearing part of the flower.

Asexual. Reproduction other than by seed. Vegetative propagation; budding, grafting as example.

Ball and burlap. Tree is dug, sold and transplanted with soil left around the roots. Burlap is commonly wrapped around the dug tree roots to keep the soil in place. Sometimes referred to as B & B.

Bare root. Trees sold without their roots in soil. The root is usually wrapped in wet sphagnum for shipping. Trees are usually dug, shipped, and planted while dormant.

Bark. External tissue layer of woody perennial plant.

Bark Slipping. Condition under which rootstock bark is easily pulled away from the tissue below. Occurs during active growth phase of the tree and is necessary when performing certain propagation techniques.

Bearing age. Age at which first blossoms and fruit are usually borne.

Biennial bearing. Production of a crop only every other year.

Biological control. Pest control by means other than synthetic chemical. Parasites, predators, or naturally occurring chemicals are usually considered biological controls.

Bitter pit. Physiological disorder caused by calcium deficiency. Appears as small, dark, round depressions on skin of affected apples.

Blush. Intermittent light red tint on the fruit skin.

Budding. Method of propagation in which a single scion bud is grafted to a rootstock piece.

Budding rubber. Small strip of rubber or plastic used to secure grafts.

Bud scale. Modified leaf or scale that serves as protective cover for an unopened bud.

Callus. Plant cell tissue overgrowth that develops in response to a wound, cut, or graft. In a graft, the callus will eventually form the graft union.

Calyx. The cup between the flower and its stem. The collective group of the sepals of an individual flower. End of the fruit opposite the stem.

Cambium. Thin layer of cell tissue between the bark and wood of a tree that is the origin of new growth.

Canker. Decayed or diseased area of the tree bark, usually exhibiting signs of gumming or oozing sap.

Canopy. The "umbrella" or above ground portion of the tree formed by the branches and leaves.

Carbohydrates. Starch, sugar, or cellulose formed by a plant.

Central leader. Single main trunk that grows vertically in the center of a tree and emerges at the top. One of several pruning systems used for fruit trees. Most used with apple.

Chilling requirement. Number of hours required below 45 degrees F in order for a fruit tree to break dormancy, grow, flower, and fruit properly.

Chlorophyll. Green pigment in the leaf that is essential for photosynthesis.

Chlorosis. A lack or loss of chlorophyll in the foliage that appears as yellowing of the leaves. Common symptom of nitrogen or other nutrient deficiency. Can also be caused by herbicide misuse.

Clay. Soil of mineral particles less than 0.002 mm in size. Has high moisture holding capacity.

Clonal Propagation. Asexual form of reproduction resulting in clone offspring.

Clone. Offspring that is genetically identical to its parent

Collar. Area where a branch grows out of the tree trunk.

Cordon. Decoratively pruned tree having several tiers of horizontally growing branches.

Crotch. Angle formed where a branch joins the main trunk or a side branch grows off a main branch.

Cross-pollination. Transfer of pollen from one flower to another.

Cultivar. Plant variety.

Deciduous. Plant that sheds its leaves at the end of each growing season.

Defoliation. Condition of having lost leaves.

Degree Days. An accumulation of heat units based on average temperatures above a given threshold.

Differentiate. Act of becoming a fruiting or a vegetative bud.

Domestic variety. Fruit varieties considered to have their origin in European or North American climate as opposed to Asian or tropical climates.

Dormancy. The condition of being dormant.

Dormant. Period during which active growth is suspended, but plant is capable of growth given proper conditions.

Dormant prune. To prune while the tree is dormant or during the dormant season (typically during the winter or very early spring).

Double leader. Occurrence of two competing, vertical growing shoots.

Drip line. Boundary of the area to which the branch tips extend. Rain drips to the ground at this boundary and forms a drip line on the ground.

Drought tolerant. Able to withstand lack of water or moisture stress conditions.

Dwarf. A tree of smaller size than a seedling would typically produce. Usually achieved by grafting to dwarfing rootstock, manipulative pruning, plant breeding, or withholding nutrients.

Erosion. Excessive washing or blowing away of soil particles.

Espalier. Decorative fruit tree trained to grow flat against a support trellis or wall.

Exoskeleton. Hard, external support covering of an insect.

Fertilization. The transfer of genetic information between male and female flower parts in the process of pollination.

Filament. Stalk that supports the anther. Male flower part.

Flower. Specialized reproductive structure.

Flower bud. Bud containing flower

parts rather than shoots.

Frass. Excretion of an insect larva.

Freeze damage. Damage to tissue by cold weather.

Frost pocket. Low lying area, prone to an accumulation of air below freezing temperature.

Fruit. Seed bearing part of the tree containing the mature ovary and related reproductive tissue.

Fruiting habit. Manner of fruiting and location of fruit on the tree.

Fruit set. Proper completion of the fertilization process, exhibited by swelling of the ovary.

Fungus. Organism with no chlorophyll, leaves, or flowers that reproduces by spores. Often responsible for fruit tree diseases. (plural: fungi).

Fungicide. Chemical used to control fungi.

Genetic dwarf. Fruit tree that, as a result of breeding, grows naturally small in size without additional manipulation.

Girdling. 1. Chewing all the way around a tree trunk by rodents. 2. Accidental constriction of branch growth by a wire or tie. 3. To sharply cut rings into the bark surface with the purpose of encouraging blossom production.

Graft Union. Area where scion and rootstock tissue are joined and grow together.

Grafting. Uniting scion and rootstock tissue to produce additional trees. see budding.

Green manure. Vegetative crop that is grown and plowed under to enhance soil.

Ground color. The base color of the fruit skin; normally green, signals harvest maturity by changing or lightening in color.

Growth habit. Natural tendency to grow in a certain shape or form (e.g.

upright vs. spreading).

Harden down. Slowing of active growth in preparation to withstand cold temperature.

Hardiness. Degree to which a plant is hardy.

Hardy. Able to withstand severe cold temperatures or to winter over without protection.

Heading back. To cut back a branch to a weakly growing lateral branch.

Hedgerow. Row of closely planted trees, grown for decorative purposes or as protection from wind.

Heel-in. To lay the tree at an angle and bury the roots in order to temporarily hold the tree if planting time is delayed.

Herbicide. Weed killer.

Honeydew. Honey consistency secretion produced by sap feeding insects.

Host Plant. A plant where an insect or disease can live.

Hybrid. Genetic cross of two plant species or varieties.

Incompatibility. Inability of scion and rootstock to form a strong graft union. Inability of pollen grain and egg to successfully form a fertilized egg that can mature.

Instar. Insect life stage that occurs between two successive molts.

Integrated Pest Management. Pest and disease control system based on understanding of fruit production as a total, interrelated cycle. Incorporates timing of manmade controls with existing natural controls to be least damaging to the natural environment. Uses selective control rather than broad spectrum controls.

Interstem. Section of rootstock grafted between scion and another rootstock, mainly to overcome incompatibility or adaptation problems.

June drop. Period approximately 15-30 days after full bloom at which poorly fertilized fruit drop from the tree.

Juvenile stage. Vegetative growth phase early in a tree's life, during which it does not produce fruit.

King bloom. First (and strongest) bloom to open in a flower cluster.

Larva. Immature wingless insect life-form that follows egg hatch. (plural: larvae).

Lateral branch. Side branch growing off of a primary scaffold branch.

Lateral bud. A bud growing on the side rather than the end of a branch.

Leaching. Washing away of minerals in the soil due to percolating water.

Leader. The most vigorous upward growing branch.

Leeward. Side sheltered from the wind.

Lenticel. Pore in the fruit skin or on the woody stem through which gases are exchanged between air and plant tissue.

Lime. Ground limestone applied to the soil to raise pH.

Loam. Soil composed of varying amounts of sand, silt, and clay.

Lopper. Long handled pruning tool used to cut branches too large in diameter for a pruning shear.

Low chill. Requiring minimal (generally less than 400 hours) exposure to temperatures below 40 degrees in order to break dormancy.

Maturity. Stage of development when fruit has achieved its highest eating or storage quality.

Metamorphosis. A marked change in physical form (as in change from a worm to a winged insect).

Microclimate. Localized area of uniform climate.

Modified leader. System of pruning in which a leader is encouraged during the tree's early life and then gradually suppressed by heading

back. Commonly used in cherry, pear and plum.

Molt. To periodically cast off an exoskeleton.

Mulch. Organic material placed on the soil to conserve soil moisture, maintain even temperature or control weeds.

Mycoplasma. A virus like organism that can cause plant diseases.

Necrosis. Death of plant tissue.

Nematode. Microscopic worm-like parasite that feeds on tree roots.

Notch. A small wedge shaped cut taken from a branch to encourage or inhibit bud break.

Nutrients. Elements necessary for plant growth.

Nymph. Immature insect stage differing mostly in size from the adult.

Open center. Pruning system where branches are pruned to a vase shape with the center of the tree remaining open. Commonly used with peaches.

Organic. 1. Containing carbon. 2. Having a natural origin as opposed to a synthetic origin.

Organic matter. Decayed leaves, roots, or wood that are part of the soil.

Ovary. Female flower part. Enlarged base of the pistil that protects the ovules.

Ovule. Female flower part. The "egg" containing the genetic nucleus.

Perennial. Plant that grows for many years, with new growth each season.

Petal. Showy, colored portion of a flower that serves to attract insects.

Petiole. The leaf stalk.

pH. Logarithmic scale from 0 to 14 that is used to express the acidity or alkalinity of the soil.

Pheromone. Insect hormone.

Photosynthesis. Process of converting water and atmospheric carbon dioxide into carbohydrates, with the help of chlorophyll in the leaves and sunlight.

Pistil. Female reproductive structure within the flower.

Pole pruner. Cutting tool on the end of a long pole, used to prune hard to reach branches.

Pollen grain. Male carrier of genetic material.

Pollenizer. Organism that assists the pollination process, usually a bee.

Pollination. Necessary part of the reproductive process where pollen is transferred from the stamen to the pistil.

Pome fruit. Fleshy fruit where several seeds are surrounded by a core.

Precocious. Tending to bear fruit at a young age.

Prune. Removal of diseased, broken or improperly located branches to maintain or improve tree health.

Pubescent. Soft, hair-like covering on the underside of leaves.

Pupa. An immature stage in the development of an insect.

Rootbound. Cramped growth of roots caused by growing in too small a container.

Rootstock. Root material onto which a productive and useful fruit variety is grafted.

Rootsucker. Shoot originating below ground from the roots or rootstock of a tree.

Rosette. Small, tight cluster of leaves growing in a bunch due to poor shoot growth. Can be symptom of boron or zinc deficiency.

Sand. Coarse textured soil particle.

Scaffold. Primary branches that arise from the tree trunk and form the main structure of the canopy.

Scion. Plant tissue grafted to the rootstock to eventually form the fruit bearing portion of the tree.

Seedling. Tree produced by growing from seed rather than by grafting.

Self-fertile. A plant able to pollinate itself and successfully produce fruit.

Self-sterile. A plant requiring cross-pollination by another variety than itself in order to successfully produce fruit.

Self-unfruitful. Inability of a blossom to be pollinated by a blossom of its own variety. Self-sterile.

Sepal. Leaf-like structure surrounding a flower bud and eventually supporting an open flower.

Shuck split. Developmental stage in stone fruit where the last of the flower is shed away from around the developing flower.

Skeletonize. To chew away leaf or bud tissue so that only the veins (or skeleton) remain.

Soil Type. Composition of the soil. Sand, loam, or clay.

Soil Texture. Fineness of the particle size of the soil type.

Split pit. Condition brought on by inconsistent soil moisture in which the pit of a stone fruit does not close properly during the pit hardening stage.

Southwest Injury. Longitudinal splitting of the tree bark in winter, usually on the southwest side of the tree, caused by uneven expansion of trunk tissue that is heated by sun reflecting off bright snow.

Spur. Modified, compactly growing branch that primarily bears fruit buds.

Stamen. Male flower part containing anther and filament.

Standard (rootstock). Size of a mature tree that is grown from a seedling of the given species. Normally used as a measure against which to express the size of a dwarf rootstock.

Stigma. Female flower part that is sticky and receptive to pollen.

Stone fruit. Fruit containing one central pit.

Style. Female flower structure that supports the stigma.

Sucker. An underground shoot arising from the rootstock.

Sunscald. see Southwest Injury.

Suture. The seam line on stone fruit.

Temperate zone. Area between the Tropic of Cancer and the North Pole that experiences annual change of warm and cold seasons.

Terminal bud. Bud at the end of a branch, that when it grows develops new shoot growth.

Thinning. 1. Removal of excess fruit from the tree. 2. Removal of excess vegetative growth to allow sunlight into the interior of the tree.

Topography. Variations in elevation of a parcel of land.

Toxicity. Poisoning of the plant by an overdose of fertilizer or sensitivity to a pesticide.

Training. To spread and position tree branches so that they develop a strong, well balanced tree structure.

Triploid. Having three times the monoploid number of chromosomes. Fruit varieties that are triploid have sterile pollen.

Variety. Group of closely related plants within the same species. A species subgroup that shares traits common to the species, but has its own individual characteristics. See cultivar.

Vegetative growth. Shoot growth.

Water logged. Soil with poor drainage, standing water, and insufficient oxygen for root growth.

Water sprout. Rapidly growing shoot that grows from latent buds in the tree branches or trunk.

Water table. The natural level of water within a given geographical area.

Whip. Newly planted tree with side branches (and often a portion of the leader) pruned off to encourage vigorous new growth.

Whorl. Collection of three or more branches radiating around the trunk.

Index

244 the Backyard Orchardist

Index 245

Index   249

250 the Backyard Orchardist

Did you borrow this book???

3 Easy Ways To Get Your Own Copy

☞ Available at local bookstores
☞ Call toll free 1-800-639-4099. VISA and Mastercard accepted.
☞ Send check or money order to the publisher:
   OttoGraphics
   8082 Maple City Rd.
   Maple City, MI 49664

Please send:

Quantity	Title	Total

_____the Backyard Orchardist:A complete guide to growing
fruit trees in the home garden @ $14.95   _____

_____the Backyard Berry Book: A hands-on guide to growing
berries, brambles, and vine fruit in the home garden @ $15.95 _____

Shipping & handling* _____
Priority/UPS _____
Foreign _____

Sales tax (6% Shipped to Michigan address) _____

Grand Total _____

Ship To:

Name: _____

Address:_____

City: _____ State: _____ Zip Code: _____

Shipping: $ 4 for the first book. $1 each additional book. Orders are normally processed within 48 hours. Books are shipped via Postal Service Book Rate. Delivery may take from 4 days to 3 weeks in the US. Priority mail/UPS shipping $6.50 per book. Foreign orders: add $4.

Satisfaction Guaranteed. If you are not satisfied with any book, you may return it, in resalable condition, along with the original invoice, within 15 days for a full refund.

Is Your Favorite Fruit Recipe a Winner?

What could be better than homegrown fruit and home cooking?

Due to popular request, we are compiling a cookbook of fruit recipes. We'd like it to be the best. Do you have a favorite dessert, preserve, main dish or other recipe that uses fruit as a major ingredient?

Could it be a real winner? To find out, why not enter it in our "Homegrown and Home Cooked" recipe contest.

Win a Complete Collection of Fruit Gardening and Cooking Supplies

Semifinalists in each category will receive a copy of the finished cookbook. The Grand Prize Winner will receive a collection of fruit growing and cooking supplies.

For details and an entry form, send a long self-addressed stamped envelope to:　　Homegrown
　　　　　　　　　　　　　OttoGraphics
　　　　　　　　　　　　　8082 Maple City Rd.
　　　　　　　　　　　　　Maple City, MI　49664

Jenny cell phone.
338-4382